PREMATURITY
THE ENIGMA
OF HUMAN
EVOLUTION

Euripedes de Aguiar

PREMATURITY

THE ENIGMA

OF HUMAN
EVOLUTION

Teresina
2018

Original title
PREMATURIDADE
O ENIGMA DA EVOLUÇÃO HUMANA

Translation
Crimson Interactive Pvt. Ltd. (Ulatus)
www.ulatus.com.br

Final technical review, cover, graphics,
drawings and layout
Euripedes de Aguiar

Impressão
Amazon's Kindle Direct Publishing

2018

Cataloguing data	
A282p	Aguiar, Euripedes de. Prematurity./Euripedes de Aguiar. Teresina, 2013. 284 p. 1. Human evolution. CDD: 576.5
ISBN-13 978-1980734925	

Original text (Portuguese), without revision, registered in Titles and Documents, at Themístocles Sampaio Notary Office, 3rd Office of Notes, Teresina-PI, on 06/03/2013, Protocol RTD No. 38185, in the form of microfilm, according to Laws 6.015 / 73 and 5,433 / 68.

1st edition registered at the Copyright Office of the National Library Foundation of the Ministry of Culture: Registration number: 630.256, Book: 1.210, Sheet: 332. Term recorded in the city of Rio de Janeiro on January 28, 2014.

email:
euripedesdeaguiar@outlook.com

There comes a moment in life
when one should risk everything even if on only a hunch.

For my brother David Machado de Aguiar
For my uncle Milton de Aguiar

ACKNOWLEDGEMENTS

To the scientists who wrote everything I read and heard,
the great database that allowed me to construct
the ideas presented in this work.
To the writers, who inspired me
with the beauty and clarity of their texts.
To the image technicians, who gave me, with their art,
the chance to reach and to elaborate new reasonings more quickly.

SUMMARY

Part IV

The past explaining the present

Part V

Raffle Theory

Cover images

Homo sapiens idaltu – Three skulls, two from adults and one from a child, were discovered in Herto Bouri (volcanic layer region) in the Middle Awash of the Afar Triangle, Ethiopia in 1997 by a team of scientists led by Tim White. Through radiometric dating, the layers of Herto Bouri were dated between 154 and 160 thousand years ago. Replica BC-045, acquired from Bone Clones, Inc., 9200 Eton Avenue Chatsworth, CA 91311 - USA, with license for commercial use of non-exclusive reciprocal photographic image. Photo licensed by Jéferson Elias Viveiros Pinheiro.

Nail monkey (*Sapajus libidinosus*) – Primate found in Brazil (Northeast and parts of the states of Bahia, Minas Gerais, Tocantins and Goiás). With an average weight of 2.9 kg, body length of approximately 41.5 cm, and tail of 43.5 cm, prefers environments with "caatinga" (dry forest) and cerrado (savanna) vegetation. It socializes in groups of around 10 individuals, though groups with as many as fifty members have been recorded. Some groups found in Serra da Capivara, Piauí use stones to chip at other stones, which they then use as tools to crush stalks and hard fruits, and to prepare sticks to hunt larvae and geckos and remove honey. Photo licensed by Jéferson Elias Viveiros Pinheiro.

Blombos Stone – The earliest evidence of geometric thinking, was discovered at the Blombos Cave in South Africa in a layer that dates to around 70,000 years ago by the team of Christopher Stuart Henshilwood, professor and researcher at the University of the Witwatersrand in Johannesburg, South Africa. Images by author Christopher Stuart Henshilwood himself, (Blombos_Cave_engrave_ochre.jpg), licensed by the Creative Commons Corporation.

PROLOGUE

"We now know that science cannot develop only from empiricism; in the constructions of science, we need free invention, which can only be confronted with experience in order to know its usefulness. This fact may have escaped previous generations, for which theoretical creation seemed to develop inductively from empiricism without the creative influence of a free construction of concepts."

Albert Einstein

In this book, I try to be sincere. It is with this intention of sincerity that I inform the readers, so that they have no doubt about what they will read, who I am, and what the main subject of this publication is. I was born in 1950, in Teresina (PI), Brazil. My higher education is in Economics, which has little to do with the subjects that will be explored here, since I present a theory to explain the emergence of human intelligence. That is, I will explain the strange emergence of humanity. It is, therefore, a new theory, which still requires criticism and proof. Several newer theories accompany this proposal to clarify phenomena regarding the evolution of humans and other living beings.

The emergence of human intelligence is one of the greatest mysteries of science, and I know that it is difficult to believe that I have the ability to embark on such a complex subject that is so different from the subjects that I formally studied. I have never participated in a single formal scientific

15

experiment or an archaeological survey, patiently breaking pebbles, looking for traces of important facts. I have never analyzed bones. I have not dissected corpses or analyzed DNA sequences. I admit it is difficult to believe that I have the intellectual prerequisites to deal with such complex issues and solve problems that humanity has never solved, and that, from what I have read, the scientific community cannot even predict when will be elucidated.

There are many scholars who believe that the mystery of the emergence of human intelligence, the object of this book, will never be properly clarified. What I can say, to try to justify my arrogance, is that I have studied these issues for more than 10 years. It is not much and doesn't begin to qualify me to even consider that I have deduced how mankind appeared. As such, I only know one way of ascertaining whether there is the slightest possibility that I am right: to write a book, to publish it and to subject myself to criticism.

Now, I think the reader understands when I say I intend to be honest. That way, if you still want to read my ideas, you certainly won't be fooled.

INTRODUCTION

I understand that the introduction of a book should prepare the reader to better understand the subjects dealt with in the text. It must have, among other things, characteristics of the work and of the author. That is what I will try to do in this chapter.

This book, certainly, is the result of a solitary exercise. The subject is vast and complex, and I had no one to help me: no teacher to consult, no former classmate to advise me, no trainee to catalogue, no one to share mistakes and corrections. But I am audacious when presenting ideas about the great mysteries of humanity: I expose what I think, without blame or apology, even knowing that I will upset many people if this publication has any editorial success.

Here within, the reader will repeatedly find expressions such as "I think," "my opinion," "my thinking," "my theory," "I believe," and so on. Unfortunately, I cannot escape these repetitions. I have come to think that a proposal like this one does not allow avoiding these common phrases without losing the clarity that I intend to achieve. Often, I am also repetitive, but apart from being a peculiarity of mine to make matters well understood, I prefer to be a bit self-evident rather than accept that my thoughts are misunderstood. Some explanations are reiterated throughout the book, especially with respect to the causal order of the emergence of human intelligence. The objective is to assure the reader while avoiding returning to the text, and at the same time illustrate reasonings with new expressions. I would rather bore those who appreciate complex texts than confuse those who expect a clearer explanation. That is why I try to write with simplicity, and I repeat when I think it necessary, to make my ideas as clear as possible, so that everyone has the right to understand them, to challenge them, to consider them or even to accept them. I prefer to lose the charm of complexity in favor of the clarity of simplicity.

The term "emergence," which appears in the title of Part 1, should generally be understood as "arising from the mechanisms of natural selection proposed by Charles Darwin and Alfred Wallace," the two naturalists who around 1858 discovered, perhaps simultaneously, the principles of the evolution of species. When I say that a property has arisen, I am not saying that it appeared suddenly, as the word usually suggests. I am saying that it is a feature that has developed through the evolution of living beings, according to the thoughts of these two scientists. Similarly, when I say that "Nature made something happen" or "produced some result," I am saying that "natural evolutionary processes have made something happen" or "have produced some result." Finally, when I speak generically about "genetic changes," or similar terminologies, I do not refer to changes in DNA alone, but to all changes that may result in the formation of living beings, both physical and behavioral.

Sometimes, I qualify animals as superior when I want to refer to animals that are evolutionarily closer to humans. Although it is an expression not advised in modern systematics, I chose to maintain it in the revision, in order to not lose the meaning of some explanations.

I used books, magazines and videos, and the Internet as my data sources, which, at the beginning of the work, was a great reference. Unfortunately, however, it has been losing this quality mainly due to the obsoleteness and even disappearance of science sites and due to the invasion of people and malicious sites trying to sell uselessness, profusely hindering the searches. Despite this shortcoming, it was of great importance in the completion of this work because most of the study and development occurred when the Internet still provided quick and objective research. Most of the books I bought were purchased on the Internet or prompted by information gained on it. To state that without the Internet, this book would not have been written is not an exaggeration.

In the **Prologue**, I referred to my proposal of a theory about the emergence of human intelligence. It is true. As I said there, I know that it is difficult to believe that I, with a limited technical training on the subjects covered, can uncover such a mystery. Even Alfred Wallace felt that the human mind was too complex to be the fruit of natural selection (Ridley, Matt. *Nature via nurture*, p. 20). However—I repeat—the only way to know if I can be right is to present my ideas for evaluation.

I also said that I try to be sincere. This is also true, even though it is difficult because humans have a tendency to modify information to substantiate their opinions. The commitment serves, at least, to keep me aware of this tendency. To fulfill this commitment to sincerity, I must first explain how I came to the main conclusion of this book and report what paths led me to conceive a theory to explain the emergence of human intelligence, which, according to me, corresponds to the emergence of humanity or, as some prefer to call it, to the emergence of modern humans.

I adopt the terminology "modern humans" to distinguish hominids that I consider to have lived before 75,000 years ago from the *Homo sapiens sapiens*. Some authors use the term *Homo sapiens sapiens* for what I call modern humans. It seems that this terminology was adopted precisely because of the cognitive difference between humans earlier than 100,000 years ago and humans after 40,000 years ago, when clear indications that a differentiated intelligence had emerged appeared in the fossil record. Thus, these authors deal with *Homo sapiens neanderthalensis* and *Homo sapiens sapiens*, which I call *Homo neanderthalensis* and modern humans, respectively, in this book. The difference occurs in the demarcation of the time when science in general says that *Homo sapiens* evolved into *Homo sapiens sapiens*, and I suggest that *Homo sapiens sapiens* evolved into modern humans. The reader will certainly see, with greater clarity, the comparison of what is generally accepted by science with what I propose in the explanations of my proposals through the course of the book.

The idea of how human intelligence would have arisen came from another theory of mine, which I consider to be much more extravagant. It seeks to explain how the general logic of the decisions of the brain of living beings works, which I published on the Internet on April 18, 2006, with very little interest from a hundred readers. Based on the comments, I do not think I was understood, perhaps due to my lack of ability to explain my theory and perhaps because the theory is truly complex and goes against common sense. In any case, I would like to thank those who have read it and commented on it. I propose this theory in Part V, so that the reader will know the initial thoughts that initiated several ideas defended in this book.

However, a question remains: Why did I come up with a theory to explain how the general decision logic of the brain of living things works? I cannot recall a special, precise reason. Between 2002 and 2004, my interest in the sciences began to increase, especially in those related to living beings. I have no explanation for this sudden interest, but the readings I undertook led me to the idea of answering a problem with a solution that would challenge the scientific community. Studying and presenting a solution, simply for the pleasure of studying and presenting a solution, was my intention. Today, I do not think that was the sole reason. Perhaps it has something to do with the assertion of age; I had reached the age of 50.

I imagined studying an organ of the human body, but I always concluded that it would be very difficult to propose something without having the equipment available to dedicated scientists in laboratories. No matter how much I tried or how many good books I acquired, I would not be able to analyze anything safely without modern equipment to prove or even test proposals and ideas. Time factor was also a hindering factor because I had a full time job and family responsibilities. Moreover, my limited English proficiency did not allow me to clearly understand technical texts from specialized areas because the best scientific books are written in English. The Portuguese translations, when they are done, are still published late. There was, therefore, no chance of success, so I put that idea aside for a while.

One day, I thought: what about the brain? Not in its physiological component because I would again have to depend on equipment, but in its logic of functioning. No one has figured out how the brain works, as I observed in an exhaustive search of the Internet. I was well aware that it was too audacious to enter into such a complex subject without further training in a subject-related course of study, to at least be taken seriously. In fact, to face such a challenge I think that I at least would require a master's degree or doctorate in a course compatible with the related subjects, such as medicine, biology, anthropology, paleontology, or neuroscience. However, without really taking myself seriously, out of curiosity rather than imagining one day publishing my conclusions, I began researching, mostly in magazines and on the Internet. I concluded that almost nothing was known about the logical processes of the functioning brain. At least, at this point, I was not at a great disadvantage compared to the scientific world, for I had an organ available for experiments: my own brain and the many other brains of people I have known in my life.

Not only people but also other animals with which, in some way, I had contact, for example, birds, dogs, and chickens. Since I was not interested in its physical functioning (neurons and synapses), I would not have to dissect any brain or require any such apparatus.

This thought, of course, is not absolutely correct. Having the opportunity to dissect brains, use modern apparatus, and read more current books would have resulted in a much better chance of success. In any case, I delved into medical books and learned that the brain is extremely complex in its physiology. I could never learn so much detail alone, especially beginning so late. The reader knows that I am not being exactly fair when I write "extremely complex." The complexity is so great that I came to think that my reasoning could never fathom. However, I wanted to understand the logic of its functioning and not how the pieces physically processed that logic. I then began to read everything that was available to me on the subject and, as I mentioned earlier, to observe my brain, those of other animals, and those of other people, requesting forgiveness for the invasion of privacy. I have found that the study of the human mind, against the backdrop of evolution, is really fascinating! Invariably, it leads to conclusions that we do not even have the courage to propose.

From there, I came to understand and explain many of my ideas about humans and various animals. At least, I think I was understanding and explaining, since I may be mistaken in my ideas, and I do not want to be a stubborn person who wants to impose his thoughts. My conclusions are really very strange and extravagant, especially those that seek to explain the logic of living beings' thoughts, about which I reach a conclusion that common sense does not accept.

I became a constant reader of the Scientific American and an aficionado of closed-channel TV shows, such as those on National Geographic, the Discovery Channel, and the History Channel, watching everything related to brains, living beings, and the formation of the universe. These channels' programming, at that time, was still focused on scientific subjects. Unfortunately, most programs today are about religion, aliens, fishing, and other subjects that I imagine attract larger audiences.

In a special edition of the journal *Viver - Mente e Cérebro (Life - Mind and Brain*, published in 2005, under license from the *Scientific American*), dedicated to memory and with the title *Memory - The physiological, neural and psychic bases of the archive of memories, I read the article "Evolutionary Records"* (CHAPOUTHIER, Georges, pp. 8 and 13). I spent days thinking about the experiments with rats described in the article. For me, something was missing to clarify the facts. It needed logic. Then, suddenly, an idea came to me clearing all my doubts and explaining everything. It was extraordinary, absurd, and beyond common sense but, in my opinion solved all unexplained issues. What was worse, despite it being difficult to accept, it suggested a simple solution. Simplicity seems to offend an important part of humanity (FOLEY, Robert. *Humans before humanity*, p. 16). I took notes, conducted tests, drew charts, and chose several opinions on the Internet; finally, I wrote an article that, based on my efforts to do it well, I thought a lot of people would understand. I published it on the Internet with the limited success I alluded to earlier in this chapter: the few who read and commented did not understand the logic; maybe, it was not clear enough.

A few months later, I took it down from the Internet. True, I was disappointed. However, I continued studying the subject, and now, I publish it here in the last chapter with minor changes to maintain the style of the book. As the theory predicts that the brain commands everything, making consciousness a mere spectator, I have acquired the involuntary habit of referring to my brain in the third person, as the attentive reader may have noted in the previous paragraph. When I can not solve a problem, I let my brain do it. As soon as it reaches a solution (if it can), it lets me know. At first, saying something like my brain can not properly relate these two databases surprised. Now, I say things like that every day.

The editorial and didactic failure of the publication of my ideas about living beings processing decisions redirected my focus. I focused my studies on humans; similar to what occurred with the theory of the decision-making logic of living beings, I published my findings. In this case, the DVD "*Evolution - The Adventure of Life*," produced by BBC and distributed in Brazil by *Editora Abril*, which I watched in early 2006 gave me the key to understanding how human intelligence came about.

A few months after having watched an animation on the DVD, I developed reasoning that seemed to resolve my doubts. I completed formulating the theory in a few days, but I have never published it. Nor did I write about it. I did not even write drafts. Even though it is simpler than the first one, I did not try to explain it in an article, for example. I did not take notes; I did not even draw a line. I only thought. I wrote while thinking and I thought while writing: all in my mind. However, the idea for a book had begun to form in my mind.

This theory had advantages over the earlier one. While my theory of the animal brain's decision-making process is based on concepts that I created and results that may contradict common sense, the theory of the emergence of human intelligence rests on widely accepted theories, proven facts, and predictions that may soon be confirmed or denied. Its results do not contradict common sense, only some universally accepted theories and concepts. Certainly, while my logic comes close to the truth in both theories, the second is much more likely to be understood and accepted.

However, one thing must be understood about the two theories: they are independent. One can be true and the other false, both true, or both false, without any contradictions.

The fact is that I deduced the theory of the emergence of human intelligence from concepts of the decision-making logic of animal's brains. However, I did not identify a single cause and effect between them. Could I have reached the second without deducing the first? I do not know; maybe someday I will have a concrete opinion on that.

An explanation is due. Would it not be better for me to call my proposals hypotheses? Maybe so, but perhaps because of the importance of the issues surrounding them, I prefer to call them theories. Since I first began to conceptualize, I have called them theories, and I have become so accustomed that changing their denotation now would disrupt my explanation. Moreover, it is in this way that most of the texts I read deal with new ideas about the evolution of living beings. Those more stickler for form can substitute "theory" for "hypothesis" throughout the book, which certainly would not compromise the work at all. I respect standards and I know their value, but on some occasions clarity and creativity prevail.

Even without writing a line about the new theory, I thought a little about it every day, and I read what I could about human evolution. It was then that, as I said earlier, when I could not find a solution to a question, I would say, "I'll leave the problem to my brain. When it solves it, it will let me know." I began to practically use the concepts of independence of the brain in relation to consciousness from my first theory. I began to notice that sometimes my brain would solve problems and let me know when it had an answer. This happens with all of us when we think we suddenly remember the solution of some problem. To put it simply, my explanation is that the brain spends days and days looking for a solution, and when it finds it, it somehow lets us know. According to my theory, it alerts the consciousness, which represents our "I," our individuality, and our concept of live human experience. I was beginning to separate consciousness from the rest of the brain and apply the theory to my life. I return to this subject at the end of the book as I try to improve my explanations of the theory on the animal brain's decision-making.

From now on, I will denominate the two theories as follows: the Raffle Theory and the Human Intelligence Emergence Theory. I use capital letters to distinguish them from similar or same word sometimes used in explanatory texts. What I call the Raffle Theory explains the animal brain's decision-making process. I claim that the brain uses a guessing process of probabilities to make decisions.

I went on to dedicate almost all my studies to human evolution, also with the intention of testing and even modifying the Human Intelligence Emergence Theory. I applied the Raffle Theory more and more to my life. With the second theory in mind, I began to consider the possibility of writing a book. It would be a simple textbook exposing my ideas, without citing sources, without references. But the more I studied, the more I became interested in writing something technical and systematic, similar to the scientific texts I was reading. The study I am now presenting is an attempt to do so. It certainly has flaws, but it was the best I could do on my own. As I said at the beginning of this introduction, I had no scientists with whom to discuss, no teachers to guide me, and no trainees to help me in this solitary exercise.

By familiarizing myself with the ideas arising from the two theories, and believing that they are correct in general, I have come to deduce and explain various properties of living things. I agree that many explanations, especially about humans, are strange and contradict legal and religious norms, precepts, and even common sense, as in the case of the Raffle Theory. But my intention is not to shock or provoke people. This book is intended only to explore science. It may not achieve this aim, but that is certainly its intention: to acquire scientific knowledge and test it by logical processes; to evaluate the scientific basis of new proposals; to humbly accept facts; and to propose theories, albeit audaciously, that contradict widely accepted theses with new visual angles engendering other interpretations.

This book has neither religious nor atheistic connotations, even though it may appear that I preach atheism. When it comes to science, where everything must be logical and proven, one cannot introduce religion, where everything depends on pure faith, without evidence. Anyone can understand that they are incompatible, but even so, there are many who seek to connect the two, especially religious people who try to reconcile some concepts of Darwin with religious explanations.

My attitude toward religions is quite simple: I am an atheist in health and religious in sickness. A better explanation would that when I am in trouble, I appeal to religion; when everything is alright, my doubts and logical reasoning return; thus, in short: I am human. I am not ashamed of this dilemma because my conclusions indicate that religious sentiment is innate to human beings. Merely a successful change gave us this aptitude. That way, when I am religious, I consider myself a Catholic because my parents and grandparents taught me Catholicism. At certain points in life, clinging to a religion is comforting and advantageous. If there are people who take advantage of this to make money, many take advantage also to make money over hunger, vanity, comfort, and even the sexual satisfaction of other people. Who knows? Perhaps it is a human trait to exploit others through faith.

When I'm an atheist, I have doubts and seek logical reason in the world in which I live. When I'm writing this book, I'm an atheist, but when my children go out on the weekends, I'm religious and pray for their safety. Religion reassures the human spirit or favors the reduction of stress-provoking brain responses; however, the reader may wish to reconsider it.

This work does not have a sexist or feminist agenda. Some of my deductions may make seem sexist because I analyze men and women differently, whereas feminism defends the equality of the sexes. However, I assert that accepting that there are behavioral differences does not necessarily say that one sex is superior or inferior to the other. Men may be more fit for some things and women for others.

It has no political agenda because, for the subject in question, politics has no interest whatsoever. To be more precise, it has less interest because politics is more a consequence than a cause here.

It has no agenda to mold the character of anybody, although in between the lines I inevitably transmit information about what I consider to be morally acceptable. After all, I am human, and humans can't resist sharing their way of life with other humans. The commitment of this book is to an explanation of facts, phenomena, and their consequences.

It has no racist agenda, nor does it deny the existence of races. But just by admitting the existence of races, I may be called a racist because currently a strong scientific current in the world believes that speaking of races is racist. I accept that there are races, but I am not a racist. It would be absurd and incoherent of me to defend the existence of a superior race that must be preserved and the others annihilated because, despite being physically akin to the white race, I have fairly reliable information that I have African and Brazilian indigenous ancestry, which is evidently descended from the far east of Asia. It would be an unspeakable stupidity to defend this policy, if we can call this kind of racism that. Regardless of which race is considered superior, I would end my life in a public crematorium or other extermination machines.

Probably because of these dogmas that systematize what is considered politically correct, few scientists dare to study racial origins so as not to be labeled racist just because they are interested in the subject, even though our understanding of how we differ from other animals is incomplete if we do not consider the process that engendered the visible differences that human groups have among themselves (DIAMOND, Jared. *The Third Chimpanzee*, p. 126). It is a complex situation. No matter how seemingly reprehensible from the point of view of racism, everything points to the fact that most humans prefer to relate to people who have the same physical characteristics.

Moreover, animals including humans, who should not have the arrogance to consider themselves different, have an aversion toward beings not resembling their relatives and even indirectly themselves. The peoples of Papua New Guinea, probably one of the groups that best represent the humans who left Africa 60,000 years ago, show extreme xenophobia (DIAMOND, Jared. *The Third Chimpanzee*, p. 15), pointing to the existence of a property that comes at least from that time. Admitting this seems to me the humblest and most correct attitude. I believe that culture is directing this aptitude to acceptable behavior, according to the most widely legalized and unfortunately unfulfilled principles of justice today.

As there is a great movement throughout the world to abolish the concept of races and replace it with that of ethnicities, I consider an explanation of how this subject will be dealt with here important. Let's start with the generally accepted concept of species: living beings with similar characteristics that cross and produce fertile offspring. Is the definition perfect? Of course not. In nature, although many people say the opposite, there is no perfection. Everything is a step away from changing, and not always for the better, depending on one's point of view. In fact, everything is constantly changing. But it is a practical definition and the best one I know.

What about races? Well, this book will consider races, or ethnicities—the term preferred by those who find that there are no behavioral inequalities between human populations—to be groups of animals of the same species that have somehow separated to acquire different physical and behavioral attributes but not enough to not produce fertile offsprings. Simple and unprejudiced. Are Caucasians and East Asians different? Of course, they are. If 10 Caucasians and 10 East Asians are grouped together, any child will know how to tell them apart. However, by crossing, they are capable of producing fertile offspring and thus are only two breeds, not two species. They are two human groups, almost certainly originating in Africa, who were for some time estranged, one in Europe and the other in the extreme east of Asia; this resulted in notable physical differences, but not so great that they can not relate sexually and produce fertile offspring. But here comes the question racists and anti-racists love because it is the great reason to debate: who is superior? I ask, superior in what? If it's to play basketball, it's Africans for sure. If it's in swimming, it's certainly Caucasians. If it's running,

27

it's certainly the African people. If it's to play volleyball, it's definitely the Caucasians. If it's to play table tennis, it is surely the Asians. If it's boxing, it's definitely the Africans. If it's to assemble electronic components, it's surely the Asians. That's how the many differences go. No one can dispute this, because the Olympics prove some of these differences every four years in an extraordinary scientific experiment that involves practically the whole of humanity. However, even in the best basketball teams in the world, we find only a few white players. There are also great white boxing champions. Or blacks who are excellent swimmers. In nature, things work out in different ways. Divisions are never absolutely perfect. Everything intertwines. But only when these characteristics differ so much that two beings are no longer able to produce fertile offspring between themselves that they will be conceptually classified as two species.

However, when racists and anti-racists argue, what they want to get to are intelligence and behavior. This is the beginning of a strange debate in which anyone who tries to study and comment on cognitive and behavioral differences between races is accused of being racist because it seems that a large part of the scientific world does not accept discussion of the subject. J. Philippe Rushton, in the abridged edition of his book *Race, Evolution, and Behavior*, defends the idea that races have far more differences than is generally accepted and invites editorial persecution from the distribution of an abridged edition that would be reaching the general public. He says that scientists know about racial differences but only accept discussion of the subject in closed scientific environments, as though the discussion could create a major social problem for humanity. At least, that is what I understood from the "Preface to the 2nd special condensed edition" of his book. How to go about this is not exactly the purpose of my work. I will not attempt to explain human racial or ethnic differences even if the reader wishes. I don't let the veiled censorship of some people deprive me of citing works of authors considered by some as racist, just as I cite the books of those considered anti-racist. That is, I do not accept censorship anywhere. But the reader who is bothered by the use of the terms "race" or "ethnicity" mentioned here can substitute one for the other. Doing so won't change the reasoning in this work because although a study of the two words meanings shows differences, practical use is turning them into synonyms in most of the works I have been reading.

It's also necessary to clarify some information. The most important is in regard to the concept of human intelligence adopted here. What is intelligence? It is the ability to solve problems. Only that? Exactly. This is the concept of intelligence that I adopt. However, by this concept even a bacterium is intelligent. It solves problems, so it's smart: that is it. A dog solves more complex problems than a bacterium can; therefore, we can place the dog on a higher level in relation to the intelligence of a bacterium. Humans solve more complex problems than a dog, so they have intelligence superior to that of a dog. Intelligence, for the understanding of this book, is a case of gradation. Darwin himself had a similar opinion: he thought that the difference between man and other animals, although large, is certainly one of degree, not of type (Ridley, Matt. *Nature via nurture*, p. 28).

In this way, I came up with a very simple and intuitive definition of human intelligence, especially in order to avoid the task, which I consider impossible for human brain patterns, to create a minimally acceptable intelligence measurement system: it is intelligence similar to human intelligence today. The reader will see that I wrote "similar", not "the same." My theory suggests that human intelligence continues to develop extremely rapidly in accordance with evolutionary levels, and human intelligence from the past will always be a little short of human intelligence today. Throughout the book, readers will be able to better assimilate my thoughts about the evolution of living beings in relation to time, especially by observing the consequences of the phenomenon known as genetic bottlenecking, which I will explain in due course. They will further realize how human intelligence may be evolving faster than the normal patterns of living evolution, and much slower than in the period in which I claim it to have established itself, between 75 and 60 thousand years ago.

Is this a vague definition that doesn't explain much? I recognize that it is. Yet it is the most understandable that I have reached in terms of my theories. It really is difficult to conceptualize the intelligence of any animal when one considers intelligence as something measured. A dog's intelligence is intelligence capable of what? Of learning answers to 20 commands? Or 40 commands, depending on the training time? Is intelligence the ability to memorize the whole course of the best places to hunt in a certain region? Of the best places that have water? What about human intelligence? Is it the ability to record many words, calculations, solutions to a certain problem? The definition of human intelligence would end up being so extensive that it would be difficult to even imagine where to store so much data in a single sentence, as it is common to do in a definition. Thus, there is no other way than to appeal to common sense and say that human intelligence is intelligence similar to that of human beings today. We human beings, who reasonably understand living things, understand more or less what human intelligence is. This understanding is enough for comprehending my theories.

Several times through the course of the book, I develop reasonings and present them as my own ideas. However, as a solo work, and therefore with a great deficiency in checking information that is already well known in scientific circles, I may be making a mistake and therefore involuntarily showing as a novelty what has already been proposed. If this proves to be true, I apologize to the author(s) for any idea that I have may have "appropriated." I will also make the necessary clarifications in case of later editions, with all the necessary identifications and information. This book seeks to be honest, and so it will be.

Every time I quote a book as a source of information, I also cite the page where it is found because I find it important to have a more complete indication of the sources on which I rely in order to develop my reasoning so that the reader is assured of the information. In light of this, even knowing the technical norms of monographs and scientific books, I choose to express a more appropriate way for me. For the reader's greater comfort, I also cite the author's first and last names, the title of the work, and the page of the book throughout the text. With regard to other sources, I give a more precise description according to necessity. The references, mentioned at the end of the book, are divided into five sources: books, articles, documents, videos, and the Internet.

For this English version, I considered removing page indications, since most of the mentioned books were originally written in English, and it would be very expensive and complicated to acquire all the mentioned English books and identify the pages from which the information originates. Even so, I thought it best to retain the page citations because, no matter how much translations may change page numbers, the information will be on a nearby page in the original version of the books. With this clarification, I believe that page number citations, even from Portuguese versions, will help the reader to more easily confirm the origins of the information. In addition, both in the course of the book and in the part concerning the books in the References, I mention the original titles, since they don't always correspond to a literal translation. I hope, therefore, to provide better comfort to readers.

When I attribute an opinion to science, I am always referring to concepts, reasonings, and theses that find greater acceptance in the scientific community, according to all the information I have accessed, including several periodical and current publications.

I call all beings descended from the ancestors of modern humans who separated from the chimpanzee around 7 million years ago "hominids". So, when I speak of hominids, I may not be referring to an ancestor of modern humans, but to a similar lineage that has become extinct. But I am certainly speaking of our very close relatives, considering the time of human evolution. Only when the explanation is essential do I use more definite terminologies such as *Ardipithecus ramidus, Australopithecus afarensis, Homo habilis,* and others.

I call modern humans with human intelligence "humans," in accordance with the definition of human intelligence that I have explained in this Introduction.

I clarify that I use the term "evolution" in the sense of the transformations that modify living beings to suit new environments, although I prefer the term "adaptation," which is what actually happens. "Evolution" has the sense of developing, improving, and perfecting, which does not seem to be the case. "Adaptation" has the sense of adjustment, which is more appropriate.

In spite of defending the term "adaptation," which is much more logical for what I imagine occurs in nature, I recognize that passing from, say, an amoeba to a fish to a dinosaur to a mammal to a primate to a human is much more akin to evolution toward development or improvement. Despite the firm understanding that this does not occur, even due to the great difficulty of determining which is more evolved than the others, in this book, I use the term "evolution" more because it seems to be better understood in the technical literature and by the readers of the subjects that I discuss.

Finally, two things: I mention the date 50,000 years ago for the complete emergence of human intelligence or humanity, as the reader may prefer, but sometimes I quote 40 or 60 thousand years apparently in the same sense.

The reader should understand that when I speak of these dates, I am always referring to a time period. Additionally, I often mention dates based on dates still under discussion. It's not at all wrong to think of 60,000 years ago as the departure of humans from Africa and 50,000 years ago as their arrival in Oceania. Either way, these inaccuracies don't interfere with the ideas outlined in this book.

Finally, I would like to clarify that I have had a difficult time confirming the data cited and the indicative guidelines, such as "before," "behind," "next chapter," etc. If any mistake has escaped my recognition, I apologize: it is a disadvantage of this solitary exercise.

Part I

The emergence
of human intelligence

1

The view of science

Most of the texts I have read about human evolution support the idea that 6 or 7 million years ago, our ancestors separated from the ancestors of chimpanzees and bonobos (or pygmy chimpanzees, as some still prefer to call them). Around 4.5 million years ago, they eventually began to use bipedal movements, a mode of displacement that they definitively adopted 2.5 million years ago, when they began the production of stone instruments, revealing a technological advance beyond the ones used by monkeys. From then on, their brains grew tremendously (KLEIN, Richard G., BLAKE, Edgar. *The Dawn of Human Culture*, p. 34). They carved stone to make tools for hunting, breaking bones, and cutting meat. For about 2.4 million years, while the brain was growing unusually, there was an insignificant increase in the quality of the tools produced. Suddenly, 40,000 years ago, and without any notable physical change, clear marks of superior intelligence began to appear on the planet. These changes were detected by the sudden appearance in the archaeological records of spearheads, drawings, paintings, sculptures, statues and adornments, stemming from a complex cognition that had never before existed, which Jared Diamond calls "The Great Leap Forward" (DIAMOND, Jared. *The Third Chimpanzee*, p. 41).

One attributes the development of this ability to make tools to the exceptional growth of the brain, which would have caused changes that prepared these hominids for the appearance of a cognition very similar to the

35

one of the humans today (DAWKINS, Richard. *The Ancestor's Tale*, p. 104). Notwithstanding the sudden appearance of signs of a different cognition than had not existed until then only 40,000 years ago, the thesis that this was the natural result of the technological advance that had begun with the advent of the first stone tools, around of 2 million years ago is widely accepted (KLEIN, Richard G., BLAKE, Edgar. *The Dawn of Human Culture*, p. 222).

Based on fossil findings and genetic studies, the authors argue that Homo sapiens sapiens developed in Africa between 200 and 70 thousand years ago, when they first migrated to Asia and spread throughout most of the planet, according to the map in Figure 1.

Thus, Homo sapiens sapiens or modern humans, as I prefer to call them in this book, left Africa around 70,000 years ago, reaching South Asia around 60,000 years ago, Australia around 50,000 years ago, Europe around 35 thousand years ago, and the Americas around 14 thousand years ago. With respect to the migration to the so-called "New World," in Figure 1, I have extended and divided the indicative arrow only to show that these humans, probably in the first millennium, continued the journey to South America, arriving at Serra da Capivara, São Raimundo Nonato, and Piauí State in the northeast of Brazil; at Santa Lagoa and State of Minas Gerais in the southeast of Brazil; and on the banks of the Chinchihuapi River in Monte Verde in central-southern Chile.

This is a somewhat simplistic synthesis of everything I've seen about the emergence of human intelligence in books, magazines, videos and the Internet since I became interested in the subject in 2001, 12 years before the first publication of this book in 2013. In fact, I found no author who referred exactly to the emergence of human intelligence. They refer generally to "how we became humans," "the first humans," "how humanity came into being," "the birth of culture," "the beginning of symbolic thinking," "What makes us human," "in search of the first man," "the lost link," "the birth of the modern mind," etc. I have adopted the expression "the emergence of human intelligence," because I think it better synthesizes all the phenomena that caused the great cognitive advance 40,000 years ago, Jared Diamond's "The Great Leap Forward," and why it leads to the discussion of what I consider to be the true singularity of modern humans: their intelligence.

The degree of their intelligence is more clearly in line with my ideas.

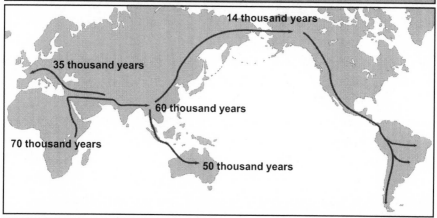

Migratory overview of *Homo sapiens sapiens* (modern humans) most accepted by science

14 thousand years

35 thousand years

60 thousand years

70 thousand years

50 thousand years

<u>70 thousand years ago:</u> for the first time, *Homo sapiens sapiens* (modern humans) leave Africa through the North of the continent, in migrations via the Middle East.

<u>60 thousand years:</u> arrival to South Asia, when they spread out to northern China.

<u>50 thousand years:</u> arrival in Australia, by sea, from island to island, taking advantage of the low sea level.

<u>35 thousand years:</u> arrival in Europe, through migrations from the Middle East and/or northern Asia.

<u>14 thousand years:</u> arrival in the Americas, via the Bering Strait, in migrations from northeast Asia.

Figure 1

Therefore, when I refer to human intelligence, I am referring only to the cognition achieved just before 40,000 years ago. This may confuse the reader, since many authors think that humankind emerged with the first stone tools 2.5 million years ago (KLEIN, Richard G., BLAKE, Edgar. *The Dawn of Human Culture*, p. 79) or with bipedalism 5.5 million years ago. In an article in *Scientific American*, Kate Wong comments on the discovery of *Sahelanthropus tchadensis* by French paleontologist Michel Brunet: "If Brunet is right, mankind may have arisen more than 1 million years earlier than estimated by a series of molecular tests." (WONG, Kate. *In Search of the First Man, Scientific American Brazil, São Paulo, Special Edition No. 2 - New Look at Human Evolution, pp. 6-15 Nov. 2003*).

37

This would lead to the appearance of humans on the planet 7 million years ago, for Brunet, although his finding was only a skull, deduced from the teeth and the base of the skull that the human was already a biped. It's all a matter of names, of course, but I'd like to clarify, so as not to confuse the reader. Many authors, especially in magazine and website articles, tend to consider all of our possible ancestors as humans after the separation of chimpanzees and bonobos, probably because these are our closest living relatives. In this view, if chimpanzees and bonobos were to become extinct, which, unfortunately, is not too unlikely, our closest living relatives would become the gorillas, and then the emergence of humanity would change to 14 million years ago when we separated from the gorillas. Chimpanzees and bonobos, like the reader, would be considered human. This doesn't make sense. The following example gives an idea of the absurdity of including them in human kind: species found up to 3 million years ago have an estimated height of 1.0 m and a brain volume of 400 cm3, whereas modern humans have an average height of 1.8 m and brain volume of 1.400 cm3. It should be noted that most of the fossils dated back 2.5 million years provoke discussion as to whether the organism was biped, quadruped or in the intermediate stage, conditions that can include it in or exclude it from human lineage. At this point in reasoning, the most important thing to know to understand my theory is that these beings did not leave the remotest trace that they had an intelligence that is similar to that of modern humans.

I believe that the 2.5 million year mark is important in the study of human evolution, especially since from then on, our ancestors completely adopted bipedalism and an erect posture. For the explanations of this book, it is not necessary that this date be precise. Therefore, it is practically certain that we had already adopted fully biped movement 2 million years ago—give or take 500,000 years. The oldest concrete proof of bipedalism, however, was the discovery in 1978. A team led by archaeologist Mary Leakey discovered in Laetoli, 45 km from the Olduvai gorge in Tanzania, the extraordinary footprints of three individuals apparently walking side by side, dating to 3.6 million years ago (DAWKINS, Richard. *The Ancestor's Tale*, p. 100). It is believed that they were made by *Australopithecus afarensis* walking on hot, damp volcanic ash. This, contrary to what may appear at first glance, does not contradict the explanation of the importance of the 2.5 million-year mark.

Knowing that biped hominids already existed 3.6 million years ago should be interpreted as the natural way of those who adopted a totally bipedal way of moving 600,000 years later.

It is necessary to add that when one speaks of the arrival of Homo sapiens in Europe, the name that designates it is Cro-Magnon because the first fossils of modern humans found in Europe were found in the shelter of the Cro-Magnon Rock, in Les Eyzies-de-Tayac, Aquitaine, Dordogne, Southeast France. In May 1868, geologist Louis Lartet found five skeletons, among them a fetus, with skulls much rounder than the Neanderthals, among other differences. During this time period, the Neanderthals were the only hominids admitted to have lived on the European continent. Although I do not often use the term "Cro-Magnon man" or simply "Cro-Magnons," preferring "modern humans" as a global reference, I am letting the reader know so that those of you who know the term won't wonder about its absence, and those who do not know it aren't confused in later readings of this book.

In Figure 2 and Figure 3, I present a comparative table of major hominids to facilitate understanding and provide an overview of human evolution. Are the hominids on this list our ancestors? Not necessarily. This is a list based on fossil discoveries. Many of them may belong to parallel lineages. In addition, fossils important for understanding human evolution have not yet been found, and many others may not even have been preserved. Still others I didn't mention exclusively for didactic reasons, especially so that a very large list of hominids does not confuse the reader. For inclusion in this list, I used a very simple criterion: those most cited in the texts I have read and the videos I have watched. However, for the reader's information, I mention some hominids that I didn't discuss: *Kenyanthropus platyops, Homo antecessor, Homo cepranensis, Homo denisovensis*, and others classified as *Erectus*, found in the Middle East, Europe and Asia: *Homo erectus lantianensis, Homo erectus nankinensis, Homo erectus palaeojavanicus, Homo erectus pekinnensis, Homo erectus soloensis, Homo erectus tautavelensis, Homo erectus yuanmouensis*. Evidently, denominations of hominids are always reasons for disagreement among scholars. But this is how science develops.

An explanation is in order. *Homo ergaster* is considered by most authors as an African *homo erectus*. However, since it is commonly agreed up on by most scientists that hominids originated in Africa, one can reason that *Erectus*

COMPARATIVE FRAMEWORK OF PRINCIPAL HOMINIDS
Information about discovered fossils

Species	Description of fossil findings
Sahelanthropus tchadensis	1 skull fossil named Toumai, Djurab Desert (Chad), 2001
Orrorin tugenensis	20 fossils of 5 individuals in the city of Tugen (Kenya), 2000.
Ardipithecus kadabba	Fossil in Affar (Ethiopia), fingers, teeth, jaws, and clavicle, 2001.
Ardipithecus ramidus	Ardi fossil, fragments of skull, arm, and pelvis, Ethiopia, 1994.
Australopithecus anamensis	Fossil discovered in Kenya, jaw, skull, and legs, 1994.
Australopithecus afarensis	Lucy (1974) and Selan (2000) fossils and others, Ethiopia.
Australopithecus bahrelghazali	Fossil named Abel, Bahr el Ghazal Valley, Chad, 1995.
Australopithecus africanus	Fossil named "Taung Child Skull," South Africa, 1924.
Australopithecus garhi	Fossils of skull and skull fragments, Ethiopia, 1996.
Paranthropus aethiopicus	Fossils in Ethiopia (1967) and Kenya, Lake Turkana (1985).
Australopithecus sediba	Fossil of woman and child, more than 220 fragments, South Africa.
Paranthropus boisei	Fossils of skull and mandible, Olduvai Gorge, Tanzania, 1959.
Homo habilis	Several fossils, Tanzania, Ethiopia, Kenya, and South Africa.
Homo rudolfensis	Fossils in Southeast and Southern Africa, (Kenya, Ethiopia, Malawi).
Paranthropus robustus	Fossils found in various regions of South Africa.
Homo georgicus	Fossils found in Dmanisi, Caucasus, Rep. Of Georgia, 2002.
Homo ergaster	"Turkana Boy" fossil, almost complete, Kenya (2002).
Homo erectus	Several fossils in Africa, Europe, Asia, Indonesia, and Oceania.
Homo heidelbergensis	Fossil jaw, near Heidelberg, Germany (1907).
Homo rhodesiensis	Fossils in Zambia, Ethiopia, Tanzania, and Algeria.
Homo neanderthalensis	Various fossils in Europe and Asia.
Homo floresienses	Hobbit; several fossils, Flores Island, Indonesia.
Homo sapiens idaltu	Fossils in Herto Bouri, Afar, Ethiopia, 1997.
Homo sapiens sapiens	Numerous fossil finds on every continent.

Note: Table drawn by the author from several sources, including the most cited information when divergences occurred. Unfortunately, I did not find a Portuguese publication with more details of the findings, especially in relation to the scientists who found them. I apologize for the omission.

Figure 2

COMPARATIVE FRAMEWORK OF PRINCIPAL HOMINIDS
Location, life span, height, and skull

Species	Location	Period of life	Height	Skull
Sahelanthropus tchadensis	Center of Africa	7.0 to 6.0 million	1.30 m	350 cm3
Orrorin tugenensis	Northeast Africa	6.0 to 5.8 million	1.20 m	-
Ardipithecus kadabba	Northeast Africa	5.8 to 5.2 million	-	-
Ardipithecus ramidus	Northeast Africa	4.8 to 4.1 million	1.20 m	330 cm3
Australopithecus anamensis	Northeast Africa	4.2 to 3.9 million	-	-
Australopithecus afarensis	Northeast Africa	4.0 to 2.7 million	1.25 m	400 cm3
Australopithecus bahrelghazali	Center of Africa	3.5 a 3.0 million	1.20 m	-
Australopithecus africanus	East/Southern Africa	3.1 a 2.0 million	1.50 m	460 cm3
Australopithecus garhi	Northeast Africa	3.0 a 2.0 million	-	450 cm3
Paranthropus aethiopicus	Northeast Africa	2.8 e 2.2 million	-	410 cm3
Australopithecus sediba	Southern Africa	2.0 a 1.8 million	1.30 m	430 cm3
Paranthropus boisei	East Africa	2.0 a 1.0 million	-	450 cm3
Homo habilis	Southeast Africa	2.4 a 1.5 million	1.40 m	650 cm3
Homo rudolfensis	Northeast Africa	2.3 a 1.6 milhões	-	700 cm3
Paranthropus robustus	Southern Africa	2.6 a 1.1 million	1.40 m	470 cm3
Homo georgicus	Northeast Europe	1.8 million	1.50 m	640 cm3
Homo ergaster	Northeast Africa	1,8 a 1.2 million	1.80 m	1.000 cm3
Homo erectus	Africa and Asia	1.8 a 0.3 million	1.50 m	1.000 cm3
Homo heidelbergensis	Africa and Europe	800–300 thousand	1.78 m	1.350 cm3
Homo rhodesiensis	Africa	600–160 thousand	-	1.300 cm3
Homo neanderthalensis	Europe and Asia	300–29 thousand	1.65 m	1.450 cm3
Homo floresienses	Oceania	15 thousand	1.00 m	380 cm3
Homo sapiens idaltu	Northeast Africa	160 thousand	-	1.450 cm3
Homo sapiens sapiens	All over the planet	200,000–present	1.80 m	1.350 cm3

Note: The data in the column "Period of life" refer to "years ago." Some data has not been filled because those fossils do not allow an evaluation. Table has been drawn by the author from several sources, including the use of averages when divergences occurred.

Figure 3

found in various parts of the world, such as China, Georgia, and Indonesia, are descendants of the first *Erectus* who left Africa or their close relatives. The reader will be able to see some of these hominids in Figure 6. However, numerous discourses persist on the classification of certain fossils, especially those found in Europe and Asia. As an example, the hominid I call *Homo georgicus* in Figure 6 is also called *Homo erectus georgicus*. There are several other hominid fossils found in Asia with this same classification problem: *Homo erectus lantianensis, Homo erectus nankinensis, Homo erectus palaeojavanicus, Homo erectus pekinnensis, Homo erectus soloensis, Homo erectus tautavelensis*, and *Homo erectus yuanmouensis*. Therefore, the reader shouldn't worry too much about the accuracy of hominid denominations. However, as I am demonstrating a theory, I have to follow the most accepted classifications and try to resolve any doubts.

These clarifications notwithstanding, the hominids that I list in the tables and graphs in this publication are quite representative of the evolutionary framework of humans, as well as sufficient for understanding the reasoning that I develop. The clues that these animals left in the form of fossils, drawings, sculptures, and others are essential for discovering how and why they broke away from chimpanzee-like beings and arrived at the humans we are. As much as we want concrete evidence in order to explain how the world in which we live arose, logical reasoning will always be necessary to fill in the gaps that chance has not preserved. To explain the emergence of life, to suggest a logical process of the functioning of the living beings' brains, and to propose a theory of the emergence of human intelligence, logical reasoning is essential. These are complex subjects that require reasoning that can even predict events that have already occurred. Often, one may commit serious errors, bordering on the ridiculous. I know that I'm running this risk, especially because I lack an academic background of the subjects discussed here.

In Figure 4, I show a chart with the same list of hominids from the previous table, relating them to their periods of existence. I have tried to obey the information I received from various sources, discarding that which showed totally divergent data. From what I have seen, they are periods obtained almost exclusively through fossil discoveries and their dating. As a result, there is a 500,000-year gap between *Ardipithecus kadabba* and *Ardipithecus ramidus*. This is certainly caused by the lack of fossil discoveries from this period.

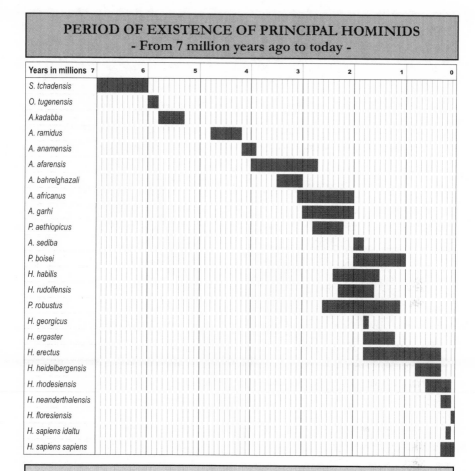

PERIOD OF EXISTENCE OF PRINCIPAL HOMINIDS
- From 7 million years ago to today -

Figure 4

The logical explanation is that either there was an unidentified intermediate hominid between the two or that one of the two had a longer lifespan—Or even that the two had longer lifetimes. Having a longer lifespan, I believe, should be recurrent in almost all the other primates mentioned in the chart. After all, they are data related to the fossils found, with information limited by several variables, such as quantity of bones, type of soil, and preservation of materials.

It is undeniable that much data is still missing in order to have a clearer picture of human evolution. The truth is that scientists don't know how

43

long each of these creatures survived (TATTERSALL, Ian. *We Were Not Alone*, 2003). Nevertheless, the data hitherto obtained corroborate a logical evolutionary process.

A more objective, simple, and didactic view compared with the realistic view of fossils is that *Homo habilis* evolved into archaic *Homo* (here encompassing various hominids, such as *Homo heidelbergensis, Homo rhodesiensis, Homo ergaster* (Africa), or *Homo erectus* (Asia), which evolved into *Homo sapiens*, into *Homo sapiens sapiens*—or as I call them in this book, modern humans. But we almost certainly came from the *Sahelanthropus tchadensis* or its close relative for *Homo sapiens sapiens*, passing through that line of hominids or their close relatives, not exactly following the order I placed them in, which depends on dates that are not always foolproof. Many dates aren't even supported by most scientists, and others are contested with new methods; for example, the datings of *Neandertais*, which modify the concept accepted to date that modern humans were encountered in Europe between 40 and 28 thousand years ago. Despite these insecurities, dating is of fundamental importance for the study of living things and particularly of human evolution, especially since the data set to date show an extraordinarily logical path supporting the main ideas of Charles Darwin and Alfred Wallace on the evolution of species.

There are, however, two conflicting theories of human evolution: (1) the multiregional one that argues that modern humans evolved independently from *Homo erectus* in various parts of the world and (2) the single origin one that holds that modern humans evolved in only one place, probably in Africa, and then spread to the rest of the world (FOLEY, Robert. *Humans Before Humanity*, pp. 129 and 131).

When comparing two hominid fossils with each other, human evolution seems strange and incoherent, but when one has an overview of all the fossils to date, the logic of the process is clear. Though complicated in its peculiarities, it is surprisingly simple in its generalities. The similarities join the differences to set up the process. Change the environment and you will feel the logic of transformations. Nothing is directed. Everything is random. Random but logical—however paradoxical that may be. This randomness of evolution, according to my understanding, does not allow an intentional direction towards humanity nor does it conform to the emergence of humanity in various places independently.

Thus, following the thinking of most of the scientists whose works I have read, I reject the multiregional theory and agree with the theory of single origin. The fossil, archaeological, and genetic evidence points to a single adaptation: a major behavioral advance that occurred in Africa about 50,000 years ago (KLEIN, Richard G., BLAKE, Edgar. *The Dawn of Human Culture*, p. 23).

2

My view
and the beginning of my proposal

In regard to the hominid fossils found to date, which I related in the previous chapter, I take the most accepted position: I dismiss *Paranthropus aethiopicus, Paranthropus boisei, Paranthropus robustus, Homo neanderthalensis*, and *Homo floresiensis*, which appear to be parallel branches. Instead, I agree with the view that we are likely to have chronologically descended from this list of fossils or their close relatives, generally accepting the established scientific thinking, which in turn obeys the principal concepts of Charles Darwin and George Wallace.

When I thought for the first time of writing a book about the emergence of human intelligence, I imagined commenting only on the events most directly related to the fact, covering a time of about 400–50 thousand years ago. I would not dare to dwell on matters before or after that. But my ideas clashed with important events, such as extraordinary brain growth. But I had no compelling explanation for this, and logic pointed to a direct cause-and-effect relationship between brain growth and the rise of human cognition. I have often thought of writing something like this: "I do not have a direct and simple explanation for the remarkable increase in the size of the brain of hominids, but I do not think it had a direct cause–and-effect relationship with the emergence of human intelligence." I would have then explained my ideas. When I began to structure the book in my mind, I understood that I could not satisfactorily demonstrate my point of view without some explanation of earlier occurrences widely accepted by science, such as remarkable brain growth, bipedalism, and the use of stone tools, in addition to the various unique behaviors of modern humans. If I thought that human intelligence had little to do with brain growth, I should then demonstrate not only how it had come about but also explain what factors had triggered such growth. Such thinking gave rise to the following question: Why has the brain grown so much? Why was this making hominids, the ancestors of humans, smarter? Was there any advantage of this increase or was there another reason(s)? To try to explain these facts and, at the same time, put the reader into the context of what I mean by evolution, I have returned to the time of the dinosaurs. However, I'm sorry to say that I haven't come up with a single, logical explanation for the brain's extraordinary growth as the relationship between increasing size and increasing intelligence. Moreover, I recognize that some behaviors remain as foreign to me as they did before I started to write the book. Better explanations will surely come, in order for us to understand more and more of how to educate modern humans to live with dignity until the vastness of eternity or as long as there is humanity.

It is now widely accepted that dinosaurs became extinct about 65 million years ago when a meteorite, comet, or large asteroid (uncertain size of 14 km to 300 km in diameter) impacted where what is today the Crater of Chicxulub, in the Mexican Yucatán peninsula, causing a catastrophe that extinguished from the planet all terrestrial species larger than 20 cm. The extinction of the vast majority of dinosaurs (the ancestors of birds escaped) ended up giving our ascendant mammals a chance to grow, develop, and dominate the earth. From then until 7 million years ago, when we separated from chimpanzees and bonobos, our ancestors were beings resembling lemurs (separated from us 63 million years ago), tarsians (58 million years ago), New World monkeys, Old World monkeys (25 million years ago), gibbons (18 million years ago), orangutans (14 million years ago), and gorillas (7 million years ago) (DAWKINS, Richard. *The Ancestor's Tale*, pp. 131-201). In this book, Dawkins considers the separation of Gorillas to have occurred 7 million years ago. However, from what I have read, now it is more accepted as 10 million years ago. He sets the separation of chimpanzees and bonobos at 6 million years ago. But it seems to me that, after the discovery of *Sahelanthropus tchadensis*, the most accepted date became 7 million years ago. However,

47

as I have already said, dating, while having a fundamental importance in the study of the evolution of living beings, shouldn't be regarded as accurate data, at least until a technology is discovered that shows more reliable dates.

I referred to "similar beings" in the previous paragraph because on the dates mentioned, the separation of our ancestors from the ancestors of beings who still live today occurred. That is, at approximately these dates there was an individual who was both our ancestor and that of the apes mentioned. For example, 18 million years ago lived a monkey as much an ancestor of modern humans as of the gibbons that now live in the forests of India, China, and Thailand. We have a natural tendency, I believe, to think that these ancestors were more like other animals than we are, although scientific theories don't indicate that. It is logical to imagine, however, that the common ancestor of modern humans and tarsians, for example, must have been a being similar to both modern humans and tarsians, just as the common ancestor of modern humans and chimpanzees must have been an animal resembling modern humans and chimpanzees. These common ancestors may not necessarily have been more like one or the other, even though this was possible. In particular, I think that our common ancestors among the various apes appeared more like them than like us because my theoretical foundations argue that we had remarkable physical changes at the onset of human intelligence.

From the extinction of the dinosaurs to our separation from the chimpanzees, two important events occurred. First, claws were replaced with nails about 60 million years ago. It is deducible that, for some animals, it was more advantageous to dig than to hunt; this animal is an ancestor of humans, who have nails, more suitable for digging, and not claws, more suitable for hunting. Moreover, after the extinction of the dinosaurs and other creatures living on the surface of the Earth, it made sense that claws, tools essential for hunting animals on land, could be replaced by nails, more appropriate tools to dig and hunt animals that lived under the earth, the more likely survivors of the catastrophe.

Second, the tail was lost 25 million years ago (DAWKINS, Richard. *The Ancestor's Tale*, pp. 149-175), which seems to me to be a consequence of its lack of usefulness. This even occurred with other animals such as bears and rabbits, apparently unrelated to cognition. There are even wild monkeys and cats with short tails, which look like they are at an intermediate stage of

48

tail loss. I believe that, about 6.5 million years ago, an animal very similar to *Sahelanthropus tchadensis* must have existed and gave origin to modern humans, chimpanzees, and bonobos, as well as to several other now-extinct species. I think this is the idea most accepted by science. My disagreement with the explanation adopted by most scientists of human evolution, I might say, begins in the motivation for brain growth, precisely because my Human Intelligence Emergence Theory does not accept a gradual emergence of differentiated cognition from the beginning of the extraordinary brain growth, as is generally accepted today. Through my studies, I have come to find explanations with which I did not agree to for other important events, such as in the case of bipedalism. I decided then to present my understandings from the beginning of the emergence of bipedal walking in our ancestors.

3

Bipedalism

I imagine that, some time after 7 million years ago, we began to leave life in the trees, spending more and more time on the ground, until we totally adopted the bipedal way of moving 2.5 million years ago. In general, bipedalism is considered an important milestone in the emergence of Homo sapiens. Truly, several authors claim that if bipedalism had not occurred, we would not have become the intelligent humans we are today. I do not agree. In my opinion, the uniqueness that makes us different from all other living things is intelligence. Do we have other singularities? We do. But other living beings also possessing them do not build cities, airplanes, and spaceships like we do. I consider that the singularity that really sets us apart from all other living creatures is the degree of our cognition. Bipedalism was only a change of locomotion most likely arising from the demand for food, as with most evolutionary changes. But since bipedalism is generally posited as one of the causes of the emergence of humanity, I will try to explain it with a new theory, while noting that other forces may have pointed to the same idea. I recognize that most theories about bipedalism do not contradict each other, but on the contrary are compatible with each other (DAWKINS, Richard, *The Ancestor's Tale*, p. 125). The following is a brief summary of some of these theories:

• Hominids who lived in the forests were forced to live on the savannas. This meant that they needed to rise up to see predators and/or the hunt better; thus, they adopted bipedalism (WONG, Kate. *In search of the first man, Scientific American Brasil, São Paulo, Special Edition # 2);*

• Hominids acquired the habit of eating while squatting, modifying the soles of the feet and causing bipedalism;

• Hominids needed to carry children and objects, and, to free their arms, they began to adopt bipedal walking—proposed by C. Owen Lovejoy of *Kent State University;*

• Hominids began to adopt bipedal walking to better regulate body temperature, exposing the body less to the scorching African sun—proposed by Peter Wheeler of John Moores University, Liverpool;

• An upright posture allowed hominids access to foods that were previously out of reach—proposed by Kevin D. Hunt;

• The need to cover large areas in order to find food in warm and open environments urged hominids to become bipedal (FOLEY, Robert. *Humans Before Humanity*, p. 233) and others.

These theories are logical and, as Dawkins comments, they do not contradict each other. I believe one or all of them may be right, since they would function together without contradicting the others.

However, none completely satisfied me. They may have contributed to the phenomenon, but the main cause was missing—something really decisive. It should be noted that the bipedal gait does not add any advantage over the quadruped. In fact, it seems to be a bad option for defense, balance and camouflage. F. Wood Jones, an English naturalist, embryologist, anatomist, and anthropologist, considered it inconceivable that a creature with such a posture could have survived without weapons (MORGAN, Elaine. *Aquatic Ape Hypothesis*, pp. 38 and 39).

I spent years thinking about it, until I came up with an idea that I found convincing. Since we are the only primates without long-projected canines, it is certain that we lost them after we separated from chimpanzees and bonobos. The reason for this loss was most probably the lack of usefulness, which leads us to believe that these hominids were living in a habitat of predominantly plant food, as they did not use their most efficient hunting weapon.

While watching several documentaries about predators, I was impressed by how important the long, projected canines are to the so-called higher

51

animals. For most of these hunters, the long, projected canines are like AR-15 rifles for modern soldiers. Just as entering a war without the AR-15 and conventional rifles today would be a disaster, hunting without the long-projected canines would also be a catastrophe for any predator in those times.

I imagine that, after a long period of abundance of plant foods, suddenly these foods became very scarce. Our ancestors then had to rely on animal food sources again. Only now they no longer had the long, projected canines.

In these cases, it seems that nature is always looking for various solutions. Most logical, of course, would be the return of the long, projected canines. But this kind of solution takes many generations, and, if it were a case of evolutionary urgency, nature would need to replace the long, projected canines with a weapon as nearly as efficient as they were, under penalty of probable species extinction. Then, I think that evolution, in the face of an animal resembling a small bonobo with some manual agility that could move over short distances, supported only on the hind legs, did what seemed faster and more logical: **it replaced the long, projecting canines with stones.**

Suppose the reader is a monkey, a baboon, for example, running on all fours and carrying stones in their hands to use as hunting weapons. There is no way to do this efficiently, and the most obvious solution is walking on the hind legs, enabling hunting with the stones in their hands (front paws), already agile due to their movement in the trees. It is possible that, in the beginning, they would just pick up the stones to attack and kill small and medium-sized animals. Then they began to use them for various functions, such as cutting the meat, breaking the bones to reach the marrow, and even as hunting projectiles.

Just like the long, projected canines, stones were also used for defense against smaller, ferocious animals and even in fights against other hominids. To this day, modern humans rely on stones to defend themselves against aggressive dogs and the like as well as—albeit unfortunately—to attack other modern humans or defend themselves. I think that, for those ancestors of ours, walking without a stone in their hands was both throwing away the chance of eventually finding a hunt and running the risk of being attacked by a predator. This seems to me a strong enough motivation for the full implementation of bipedalism.

By this reasoning, I consider that the loss of the long, projected canines, coupled with a radical change over a certain period in the vegetarian diet in favor of a carnivorous one, was the major cause of the emergence of bipedalism causing the replacement of the long, projected canines with the use of stones. The use of stones required the freeing of the hands to carry and handle them, forcing a new way of locomotion using only the hind legs. Moreover, there is no sign that, once having descended from the trees, they have used their hands to move about on the ground like chimpanzees, bonobos, gorillas, and baboons continue to do. It is as if they descended with their hands already occupied, and then they simply got up to walk (MORGAN, Elaine. *The Aquatic Ape Hypothesis*, p. 37). Even though I'm not entirely sure, I think my proposal can join the others, since it doesn't directly contradict any of them.

4

The cry of human babies

Further on, after presenting my theory for the emergence of human intelligence, I carry out an analysis of various physical and behavioral characteristics of modern humans, trying to identify their origins in light of my ideas. However, I will comment on some of them whenever I consider it important in the context of the explanations. This is what I will do now in regard to the crying of our babies.

I have always been intrigued by this attribute of modern humans, which, in principle, mainly due to the strident and extraordinary noise provoked, would only serve to attract predators. I think it is even a rarity among so-called superior animals, only paralleled in dogs, a phenomenon that I will try to explain later in this chapter. In the absence of a more convincing motivation, I came to consider that the crying of human babies would have appeared after the emergence of human intelligence, when modern humans were already living in large numbers, and thus more protected from predators that might have been attracted by the howling noise of our weeping offspring.

The books and articles I've read and the documentaries I've seen do not mention this property. Therefore, I'm not aware of any scientific justification for the phenomenon. The explanation I have often heard from most people is that human babies cry because they are hungry or because they are feeling some discomfort, such as pain. Although this is a logical answer, I have never been completely convinced of this interpretation. I've always looked for something more plausible, until, in elaborating my point of view on the emergence of bipedalism in hominids, I came up with a much more reasonable proposition for the crying of human babies, reinforcing the adoption of bipedal walking.

All primates, except modern humans, are born with the ability to cling to their mothers. This probably has to do with the fact that most primates have very long, strong arms to move quickly through the trees. Our ancestral hominids, descending from the trees to an eminently bipedal and terrestrial life, reduced the size of their arms and lost the strength and ability to move among the branches. This had an effect on the babies, who lost their ability to cling to their mothers and therefore became completely dependent. It is easy to imagine that for a hominid baby living in an environment full of predators, from cats to snakes, from canines to insects, staying on the ground even for a short time would have been extremely risky. By this reasoning, such a radical and dangerous pattern as loud, childish crying might have arisen, not for the baby to signal hunger or pain, but to protest when its mother pushed it out of the protected environment of her lap.

My proposal, therefore, is that human babies essentially don't cry because they are hungry or in pain, but because they want to be in the safest place for them in that jungle of beasts that was Africa of those times. It's as if they said, "Mother, if you leave me here on the floor for another minute, I will scream and call on the lion to eat you." Despite this explanation, I consider that babies too, by associating the discomfort of danger with the suffering of hunger and pain, began to use crying to also convey hunger or pain.

Thus, by my reasoning, the carrying of babies in human arms would have served as an auxiliary force, especially in relation to females, of the emergence of bipedalism, since I propose that the main cause was the replacement of the long, projected canines by stones. I note that C. Owen Lovejoy reinforced the thesis that the need to carry babies in arms influenced the implantation of bipedalism.

The crying of babies would therefore be a direct consequence of the descent from the trees, which also caused a decrease in the length and strength of the arms of human babies.

Now, I need to justify the crying of puppies because, after all, it is a property very similar to that of modern humans. I will attempt an explanation without much support in facts, since I have no in-depth knowledge about dogs, wolves, coyotes, or foxes. As far as I know, the babies of wolves, coyotes, and foxes, the closest relatives of dogs, don't have the strident cry of puppies, which seems quite consistent since such crying would certainly attract predators. Therefore, it's plausible to believe that this behavior in puppies has appeared recently, after a species of wolves approached modern humans and were domesticated, originating the modern dog (DAWKINS, Richard. *The Ancestor's Tale*, p. 53). Consequently, it is reasonable to think that modern humans have something to do with it.

I've raised dogs for a long time. I have followed the birth of hundreds of them, and I have observed a curious fact: puppies like humans more than their own mothers. When they open their eyes, a few days after birth, puppies begin to devote their attention and affection to humans, leaving only the task of feeding to the mother. They fill their bellies, move away from their mothers and follow after the humans. It seems absurd, and I have never seen this information anywhere, but with every birth of a litter, the result is always the same.

When and why do puppies cry? They cry mainly when they wake up and their mother is not with them. When the mother arrives, they stop crying. Thus, my conclusion is that puppies cry to call their mothers when they are hungry. In an environment protected from predators by modern humans, there is no danger in alerting them. But I presume also that they cry in order to advise humans that they, the puppies, are in danger because the mother is not with them.

I performed the experiment described below a dozen times, invariably with the same result. After the puppies have nursed and are sleeping on a full stomach, I take the mother away. When they wake up, they immediately begin to cry. Although sonorously different, it is as irritating as that of human babies. The two cries seem to be made to annoy the ears of modern humans. I think this is correct because it is enough that I approach for them to stop crying immediately, wag their tails, and curl around my legs. It's as if the cry had the double meaning of calling the mother to feed them and to humans to take action. From then on, puppies move further and further away from the mother and come closer to humans, even differentiating among them, until they appear to not even know their own mothers, often violently attacking them. Some adult dogs, when they are confined, cry like puppies until they are released by humans. However, let us return to the object of focus—ancestors of modern humans.

Having explained the replacement of long, projected canines with stones and the crying of hominid babies, I relate some events that happened between 7 and 2.5 million years ago, and the attentive reader will of course notice that I have not explained anything about the prolongation our ancestors' infancy. I will do so in the next chapter.

FIRST LIST (Part I)
Beginning of the period: 7 million years ago
- separation of human ancestors from chimpanzees and bonobos;
- loss of long projection canines (ea);
- replacement of long projecting canines with stones (ea);
- increased use of bipedal walking;
- adoption of fully biped locomotion;
- first extension of childhood;
- emergence of the crying of babies (ea).
End of the period: 2.5 million years ago

Later in Chapter 6, I explain the insertion of my initials, (ea), after listing some proposed events.

One thing is absolutely certain: the infancy of modern humans is much longer than that of chimpanzees and bonobos, our closest relatives. Most scientists I read about defend the idea that there was a prolongation of modern human infancy, not a diminution of chimpanzees' and bonobos' infancies. I estimate that an animal that has to learn to walk as we humans walk today would need a prolonged childhood for learning this vital skill. It was for this reason that I placed this event in that period. However, I have found that there was a greater growth in childhood at the onset of human intelligence for reasons more closely related to the complete development of the new brain with a differentiated cognition, according to a second list of events that I will present later in Chapter 6.

The occurrence or not of this episode in this period has no causal relation with my Human Intelligence Emergence Theory. I am assuming events that seem logical to me, in a general context, for the reader to judge. Only for this reason—the need to learn to walk as a consequence of the new mode of locomotion—could a prolongation from childhood for learning have occurred. I think that, in a small way, the demand for a slightly larger brain came from this period. This larger brain was needed to house a complex nervous system capable of controlling an extraordinary and complex network of nerves directly connected to the muscles and to ensure the necessary adjustment to the balance of singular and extraordinary human bipedal locomotion.

5

Three thoughts

I propose that the ancestors of modern humans came 2.5 million years ago to a fully bipedal locomotion, without long, projecting canines but with a weapon that seemed to perform the same function: stones. They also had babies crying unbearably. From now on, the reader should expect a greater confrontation between what is widely accepted by science and my ideas, usually manifested in the form of theories. I will try, nevertheless, to always clarify what is accepted by science and what I am suggesting. I confidently hope to accomplish this task without committing any injustice.

I chose not to specify the dates of the events, preferring order, so that the understanding of my view of the evolution of hominids is as clear as possible. The order of the lists in this way is more in the didactic than in the chronological sense. Events that I place in a certain order may well have occurred at the same time or even in a different order, depending on the consequences they may carry for general understanding. The emergence of the loud crying of babies, for example, may well have begun before the adoption of a fully biped locomotion without contradicting for reasonings I present. Despite these caveats, I try to follow a chronological order in line with my ideas whenever possible.

I also do not assume that these events occurred exactly one after another. I believe that some happened at the same time, perhaps one starting while the other had not yet finished. Even the limits of 7 and 2.5 million years are more didactic than actual or supposedly proven. The separation of chimpanzees may have been a million years before or after. The complete realization of bipedalism may have been 3.5–1.5 million years ago. It is important to note that the imprecision of these dates does not hinder, at least in principle, the explanation of the Human Intelligence Emergence Theory. However, it should not be forgotten that these inaccuracies have a limit within which they may simply preclude some theories. The study of these boundaries will certainly be done in this book to demonstrate the viability of my ideas.

If bipedalism was the great event of the first 4.5 million years after the separation of chimpanzees and bonobos, the extraordinary growth of the brain took place in the next 2 million years. In almost all the texts I have read, these two events in some way compete to define the emergence of humanity, one or another assuming greater importance. Bipedalism has a certain advantage because it occurred before and, therefore, it is often pointed out as the determining factor of the emergence of humanity. But the extraordinary growth of the brain has its strong point in a direct relation to human cognition, a recognized singularity. Following this tendency to value these two phenomena, I have chosen three thoughts that most directly indicate the emergence of human intelligence, when attempting to explain some singularities that made us human, starting 7 million years ago.

These are creative, original, and even extravagant ideas about bipedalism and brain growth. Their authors are Richard Wrangham, Elaine Morgan and Richard Dawkins. But, truly, they do not propose a theory to explain how our intelligence differed from the intelligences of other living beings. They show only causes for the emergence of bipedalism and brain growth, which they consider moreover, as is widely accepted by most scientists, the determining facts that have resulted in the emergence of extraordinary human cognition. I therefore ask permission from these scientists to attempt to explain their ideas as a kind of preparation for the reader to better understand my arguments.

Richard Wrangham says that we learned to cook, and, because of this, we acquired the prerequisites for the emergence of human cognition. He sets forth a new theory that defends the idea that cooking has made us human. Wrangham considers *habilines*, which is what he calls *Australopithecus habilis* or *Homo habilis*, the precursory species of human intelligence, maintaining that the onset of the emergence of mankind occurred 2.3 million years ago (WRANGHAM, *Richard. Catching Fire—How Cooking Made Us Human*, p. 9). He stresses that cooking allowed us to open, cut, and grind hard foods, but, above all, it significantly increased the amount of energy our bodies could get from food. He thinks that this extra energy brought several advantages to the first cooks, with positive changes in anatomy, physiology, ecology, psychology, and socialization. Above all, it ensured the energy supply necessary for brain growth.

Elaine Morgan defends the idea that, 5 million years ago, our ancestors had a semiaquatic life before returning to a predominantly terrestrial life, and this would have caused several anatomical consequences that flowed into the emergence of humanity. However, this theory is not originally Morgan's. It is from Max Westenhofer, a German scientist who defended his thinking in the book *Das Eigenweg des Menschen*, published in 1942 in Berlin. It was not highly considered. It really was very extravagant for the time, and even for the present day. In 1960, the American scientist Alister Hardy defended the same thought with new proposals and explanations (MORGAN, Elaine. *The Aquatic Ape Hypothesis*, pp. 18 and 148). It is to Elaine Morgan's merit that she has supported the thesis over the last few years.

For her, having a semiaquatic life caused a process that directed our ancestors to become the humans we are today, with several characteristics that differentiate us from the other primates, such as the thinning of hair, remarkable subcutaneous fat, a turned-down nose, and a larynx, among others. Richard Wrangham suggests that our ancestors learned to cook by increasing the use of proteins necessary for the growth and development of the brain and then began a process that brought them to intelligence and made them human. Evidently, they show causes, but they don't speak of an objective with regard to the transformation of a being with a similar intelligence to that of a chimpanzee into another with extraordinary human cognitive qualities. I read their books very carefully. I learned much from them, but I don't agree with their ideas and conclusions at all. . The focus of the ideas is excellent and full of information, which I used to reinforce, and even change my arguments.

In contrast, Richard Dawkins despite clearly stating that he is suggesting a theory to explain the increase in human brain size, develops a rationale that, in my view, points to the emergence of human intelligence, perhaps because the emergence of human intelligence is generally accepted by science to be a consequence of brain growth. I first encountered Dawkins' view of brain growth in late 2009 when I read his book *The Ancestor's Tale*, in which he says his understanding is best developed in the last chapter of his other book, *Unweaving the Rainbow*. I was able to acquire and read its last chapter at the beginning of 2012. In both books, Dawkins described a creative analogy of the human brain with computers because he thinks that brain growth, because it has been inflationary, deserves an inflationary explanation. He calls his theory "software and hardware co-evolution," and says that software and hardware innovations drive one another in a growing spiral (DAWKINS, Richard. *The Ancestor's Tale*, p. 114). The software he considers as candidates for propelling inflationary brain growth are language, animal tracking, object throwing, and memes.

Most readers, I think, are unfamiliar with the meme concept. Therefore, I'm going to open a parenthesis for an explanation. A meme, a term coined in 1976 by Richard Dawkins himself, in his first book, *The Selfish Gene*, is a unit of culture that can be understood separately and is capable of replicating from brain to brain. Examples of memes are melodies, parts of melodies, ideas, pieces of ideas, slogans, clothing fashions, and ways of making pots or building bows. However, in this book I don't use that term, preferring to use that of "culture" in the sense that scientists adopt it, that is, as the knowledge and behavior that living beings conceive of or receive from other living things. In such a definition, memes will be included. Later on, I conceptualize culture and instinct side by side, so that the reader can better understand the position adopted here.

Dawkins thinks that new software, for example, language, when it emerged, changed the environment in which brain hardware was subject to natural selection, giving rise to strong Darwinian pressure to perfect and increase hardware in order to take advantage of the new software, and a self-powered spiral started to work with explosive results in increasing brain size.

I ask, with the respect of one who have learned much from his books, that Dawkins allow me to disagree with this reasoning, that the emergence of software can run a self-feeding spiral without external factors that cause the necessary physical changes. As an example, I'm going to use the first candidate Dawkins used for software: human language. I write human language because I consider that the vast majority of animals, if not all, have a language, only much simpler than sophisticated human language.

Many authors associate language with human intelligence (DIAMOND, Jared. *The Third Chimpanzee*, pp. 159, 160, and 390). I agree, in part, with the association, but I don't think that language was the cause of the emergence of human intelligence because language as a means of communication is shared by almost all living beings, each in its own way. Monkeys, lions, canaries, chimpanzees, gorillas, dogs, and chickens all have their language. My point of view is that before human intelligence, we had a language compatible with the animal we were. But, as we began to develop an intelligence similar to that of modern humans, we began to develop a language compatible with the new cognitive system, that is, with the new type of animal we were becoming. Language, as with almost everything that relates to animal intelligence, can be scaled, even if imprecisely.

I raised chickens as a child. They were a fighting breed, called fighting cocks. I spent hours in the chicken coop taking care of chicks, laying hens, breeding chickens and roosters, and training the cocks for fighting. At that time, as far as I know, this wasn't a crime. The result: I learned their language. I remember to this day at least seven types of calls that meant different messages: the rooster clucking to indicate he was the owner of the place, which, with some logic, some rare chickens emit when they win a fight with another chicken; the clucking of the hen indicating her search for a place to lay egg, which, also with some logic, is the same as the rooster's when he is suffering injury after losing a fight; the clucking of the hen to announce that she has hatched her eggs, which strangely is the same as the rooster's when he seems preoccupied with its brood's safety; a warning call about danger the brood, exercised by the cock, the hens, and even the chicks, clearly demonstrating an instinctive language; a warning call indicating that the bird is willing to fight, exercised by roosters, hens, and chicks; the call to eat or convey that food has been found, exercised by both the rooster and the hen; and the short and repetitive chorus of the hatched chicks and the mother hen, used to not lose each other.

Those are the ones I remember. The call to eat is interesting. Roosters and hens carry it out for different purposes. Hens usually use it to call chicks when they find some food, and roosters when they find some food and want to offer it to the hens. The example of chickens applies to almost all living beings, which somehow communicate, and human language differs from others only by the capacity of the system that produced it: human cognition.

I won't resist giving my opinion about the role of language in human evolution, putting the cart a little in front of the horse. For me, human language did not produce human intelligence, but rather the opposite: human intelligence is what created human language. Earlier hominids also had a language. But it was much simpler than human language. Human language, in my opinion, is the result of a cognitive system capable of processing it. If evolution had produced a human intelligence far superior to the one it actually produced, I suppose we could have a much more powerful, sophisticated, and complex language, and, I daresay, imagining that our cognitive system could at least understand it, it would probably be much more beautiful.

I believe that by using my explanation of language, we can better understand Dawkins' argument. Imagine a hominid with a small brain and a vocabulary equivalent to that of a chimpanzee. Of what use would software similar to that of human language, able to work with thousands of pieces of information such as words and rules of syntax, be in the brain of this hominid, if in that brain there wasn't room to record such amounts of data? I think that if, by a chance - that I don't believe - software such as human language arose in a brain analogous to a chimpanzee's, that software would have been lost in evolution due to an absolute lack of utility as it would exist in a brain without sufficient information for it to be used nor space to record it.

My thinking, if correct, does not in any way invalidate Dawkins' proposed software/hardware self-feeding cycle. It merely demonstrates that there is a limit to this without an external factor, which I believe, is brought about by the immediate advantages, but with mid- and long-term results, of feeding/breeding. Within that reasoning, especially considering the time span between the generations of hominids and the number of individuals, I do not see how the software/hardware self-spiralling can work, especially when one considers that this has not happened to any other animal in the millions of years of evolution.

Similarly, if there were to be found in the brain of this hominid a space to store enough data for the operation of software such as human language, that is, a hardware with greater capacity to store data, it would be absolutely unlikely that it would also appear in the next generations. A software capable of using all that space would end up being lost in the process of natural selection, also for lack of utility. Nature is usually economical: it does not entertain costly, useless novelties. To expect an increase, even if gradual, in the processing of software (language) and hardware (data space), one feeding the other is, in my view, forcing an intentionality that the evolution of living beings does not have. In other words, I think that the evolution of living beings has come to admit some intentionality, even if it is minimal.

I had much sympathy for Dawkins' idea when I read it for the first time, mainly because it draws an analogy with computers, just as I do in the explanations of my theory for the emergence of human intelligence. But the differences from my convictions are so great that, given the author's importance and the great respect I have for his ideas, I have spent time imagining that I shouldn't write this book. I knew that I could expose myself to ridicule since I would contest an almost unanimous scientific thought that human intelligence arose because of brain growth.

However, it was good that this happened because I studied with much more dedication, and slowly, I regained my self-confidence. However, I confess that I still have some fear of being completely wrong today and that this may somehow embarrass me. I deem it to be a natural fear, and the result, above all, of two factors: the extravagance of my ideas and the fact that I studied and wrote alone. As a counterweight, perhaps because I'm not an expert on the subject, I have a greater audacity to develop and propose new ideas. In any case, the curiosity to know if I am at least right in some aspects pushed me to complete and publish this book. I think I did the best I could.

It should be noted that Dawkins, apart from not presenting the theory of "software and hardware coevolution" as a solution to the emergence of human intelligence, admits that brain growth may have been a result of the sexual selection mentioned by Charles Darwin in his book *The Offspring of Man and Selection in Relation to Sex*, in 1871 (DAWKINS, Richard. *The Ancestor's Tale*, pp. 114, 315 and 316). It is the reasoning that Darwin used to explain the male peacock's tail, seemingly of no use at all, and even considered an encumbrance. For Darwin, female peacocks appreciate males with pretty tails, and that's enough to provoke a chain reaction, making the appearance of males and the taste of females evolve together. I prefer the logic of Alfred Russel Wallace, who, simultaneously with Charles Darwin, discovered natural selection and who disagreed with the arbitrariness of Darwinian sexual selection (DAWKINS, Richard. *The Ancestor's Tale*, p. 317). He explains that female peacocks choose males not by whim, but by merit. In this way, peacock males with the most beautiful tails would have the best genetics, and females who liked males with beautiful tails would have offspring who were more likely to survive and reproduce. It makes sense and is more coherent, but in theory, the two opinions may explain the sudden increase in brain size. However, I prefer not to individualize sexual selection much. I choose to view it only as a part of natural selection. In my thinking, a peacock female would choose a pretty-tailed male in order to have healthier children in the same way that, if it were the case of choosing between two children to survive, as the dogs do choose the one with the most healthy signs. Beautiful tails would be the expression of "good genetics and thus of ideal parents. While agreeing with Wallace's understanding of sexual selection, I do not think it can by itself unleash and maintain a self-sustained software/hardware spiral that would account for the inflationary increase of the brain.

Although I disagree with their conclusions, I agree with several proposals by Dawkins, Morgan, and Wrangham, and even incorporate some of their ideas in the formatting and explanation of my theory as the reader may observe in the following chapters.

6

The extraordinary
growth of the brain

The most accepted reasoning for human evolution always considers an apparently inevitable trajectory, beginning around 4.5 million years ago with the onset of bipedalism and continuing with extraordinary brain growth around 2.5 million years ago (MORGAN, Elaine. *The Aquatic Ape Hypothesis*, p. 138). It really is logical reasoning, with several proven facts supporting it. When I read the first few texts on this subject, I agreed that a natural increase in cognition was occurring, caused by an increase in the size of the brain; I was influenced both by the texts I had read as well a by the facts themselves, which I admit are, or at least appear to be, very rational. Besides everything appearing coherent, this thesis differentiates us from other living beings, satisfying the ego of the most important species on the planet. The reader can get a glimpse into the logic of the evolution of the hominid brain in Figure 5, which I elaborated on the basis of fossil findings.

A growing brain and increasingly sophisticated artifacts—even if this increase in sophistication is minimal—form a picture of a fulminating logic. This seemed irrefutable: the brain grows, and intelligence grows along with it. Problem solved. After all, almost all of what we mean by intelligence takes place in the brain. I believe that the realization that the appearance of bipedal walking and brain growth happened in parallel to or at least sequenced with the appearance of the first tools associated these two facts with the emergence of humanity (DIAMOND, Jared. *The Third Chimpanzee*, pp. 20, 21, 46 and 47).

EVOLUTION OF THE CRANIUM OF HOMINIDS

1.450 cm3   **Homo neanderthalensis**

1.450 cm3   **Homo sapiens idaltu**

1.350 cm3   **Homo sapiens sapiens**

Homo erectus   1.000 cm3

Homo habilis   650 cm3

Australopitecos africanos   460 cm3

Australopitecos afarensis   400 cm3

Chimpanzees   400 cm3

I have drawn this illustration, copying several versions of skulls of hominids and chimpanzees, and attempted to combine the information of photos as well as drawings available to me. I have also tried to be as faithful to proportions as possible so that the reader has a complete view of how the hominid brain has evolved. Thus, I thank the authors of the photos and drawings that inspired this work.

Figure 5

Despite this, over time I began to grow concerned about some questions that this logic does not answer: the hominid brain grew extraordinarily for 2.4 million years, from 400 cm3 to 1,400 cm3, but the tools produced by these hominids changed very little over the same period. The pace of technological change during the span from 2 million to about 300,000 years ago is surprisingly slow (FOLEY, Robert. *Humans before Humanity*, p. 252). In that period, practically only one tool was created, the hand ax—initially a minimally worked stone. As the brain tripled in volume, the tools became only slightly more elaborate.

Even Homo sapiens have not shown to possess modern technology for more than 60,000 years (FOLEY, Robert. *Humans Before Humanity*, p. 252). Why, with no remarkable brain growth, did hominids (now called humans) or modern humans begin to show an extraordinary problem-solving capacity 40,000 years ago, demonstrated by the production of spearheads, drawings, paintings, sculptures, statuettes, and ornaments? (DAWKINS, Richard. *The Ancestor's Tale*, pp. 55 and 56)

Why did the Neanderthals, who lived in Europe and parts of Asia from 300,000 to 28,000 years ago, have a larger brain than modern humans (around 10%, according to most of the texts I have read) but did not have at least the same cognitive capacity and were extinct when modern humans came to Europe around 35,000 years ago? It should be noted that Neanderthals were much stronger than modern humans, and they were in their own habitat. Despite this, they became extinct in just over 10,000 years of coexistence, probably 28,000 years ago (DAWKINS, Richard. *The Ancestor's Tale*, pp. 88 and 89). The answer seems to be that Neanderthals, even with larger brains, did not have modern humans' cognitive ability and could not compete with them. It is logical to think that these hominids who were coming to Europe had something that no other animal had or no living creature had had until then. Archeology thus demonstrates no clear signs of intelligence were ever found in Europe before the arrival of modern humans on the continent (KLEIN, Richard G., BLAKE, Edgar. *The Dawn of Human Culture*, p. 149).

Even if one accepts the general opinion of science today that human intelligence began to be delineated 2.5 million years ago with brain growth, logic points out that just before 40,000 years ago something very important happened with the human beings, who had demonstrated archeological records of spears, drawings, paintings, sculptures, statues, and adornments, which certainly originated from a complex cognition never existing on the planet and very much like that of today's humanity.

In this respect, most authors, even though human cognition had begun much earlier with brain growth and the production of stone tools, agree that around 40 or 45 thousand years ago something extraordinary happened with humans that caused so many formidable developments (DIAMOND, Jared. *The Third Chimpanzee*, pp. 41 and 390).

Consequently, one can say with very little possibility of error that, betweeb 2.5 million years and 40,000 years ago, something must have occurred that provoked the emergence of human intelligence. In other words, several events eventually provoked, somewhere between 2.5 million years and 40 thousand years ago, the emergence of humans with a cognition very similar to ours.

Additionally, in this period, there were several changes in the human ancestors, which I relate below:

SECOND LIST (Part I)
Beginning of the period: 2.5 million years ago
- large increase in brain size;
- appreciable improvement in balance in the bipedal stage (ea);
- increased accuracy in throwing objects (ea);
- enlargement of buttocks (ea);
- great increase in height;
- growth of the penis (ea);
- manufacture of stone tools;
- small increase in the quality of stone tools;
- small reduction in opening of the pelvis (ea);
- small reduction in brain size (ea);
- decreased gestation period (ea);
- second prolongation of childhood (ea);
- change in nose shape, with downward pointing nostrils (in);
- thinning of body hair (in);
- appearance of a layer of subcutaneous fat (in);
- turning of the lips (in);

- control of respiratory movements (in);
- ability to also breathe through the mouth (in);
- laryngeal decay (in);
- perspiration as a way of controlling body temperature (in);
- fire control (ea);
- cooking food (rw);
- sense of immortality (ea);
- idea of God (ea);
- female orgasm (ea);
- occult ovulation (ea);
- constant sexual receptivity (ea);
- menopause (ea);
- increase in lifespan (ea);
- monogamy (ea);
- mating preferences (ea);
- male and female adultery (ea);
- private sex (ea);
- male and female homosexuality (ea);
- African genetic diversity (ea);
- Modern humans migrate to Asia and Oceania (ea);
- Clear marks of human intelligence detected through the sudden appearance in the archaeological records of spearheads, drawings, paintings, sculptures, statuettes, and ornaments.

End of the period: 50 thousand years ago

The date of the end of the period, 50,000 years ago, has a more didactic connotation. In fact, its definition will always depend on evidence dating the existence of humans with intelligence similar to the intelligence of humans today.

By the way, an interesting fact deserves to be mentioned at this point: all known species of hominids dating back more than 2 million years ago are uniquely African, more specifically sub-Saharan. (FOLEY, Robert. *Humans before Humanity*, p. 140). This means that fossil finds show that our ancestors or their close relatives only left the African continent 2 million years ago. It is very important to note that this also means that hominids not yet provided with an intelligence like ours arrived in Asia and Europe before the intelligent hominids, who only spread around the world around 60,000. Through Figure 6, the reader may have a clearer view of how hominids, who emerged after 7 million years ago when the branch that gave birth to chimpanzees appeared, lived and evolved on the African continent and migrated to Asia and Europe only 2 million years ago.

Since I list events that have not been quoted in any source I have access to, or at least in the contexts in which I include them, I identify them with my initials in parentheses (ea) to clarify that I take responsibility for them or interpretations that I make from them. I intend, in this way, to prevent the reader from understanding them as scientifically accepted facts. I also identify food cooking, the basis of Richard Wrangham's theory with his initials

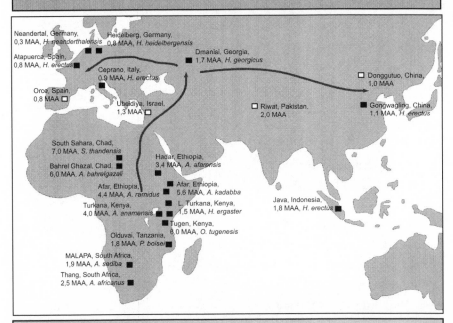

Figure 6

(rw), and the events that Elaine Morgan relates in her book *The Aquatic Ape Hypothesis* with her initials (em) because in general they are rarely cited in other publications. In doing so, I hope to respect their works and their ideas, as well as express my admiration and gratitude. This, of course, does not mean that I am claiming that these scientists and I are the identifiers of these events. It may even should be that someone has already identified some of them. But I, with my linguistic limitations, could not access that information. Therefore, my commitment to sincerity is still valid, as well as my commitment to the authorship of proposals, ideas, and theories. I reiterate, if I take some ideas in some part of this book that somebody has already published, I apologize, and I will certainly make the necessary corrections and considerations in later editions.

However, here, more than on the list I drew for the period from 7 to 2.5 million years ago, the order of events may be different from what I suggest, most of the time without prejudice to logic, rationality, and explanations. As already mentioned, the order shown is more of a didactic question because there is no chronological commitment that is precisely exact. In some cases, I do not even have a formed opinion about when a fact occurred in relation to others, and its order on the list is more a guess than a logical conclusion. In fact, many events may have occurred together, or in reverse, or some completing while others continued to unfold. To assemble the complete mosaic of evolution will always be a dream of those who study the subject. Improving knowledge of how it happened, however, will always be a fascinating activity, which will be developed through study, research, and proposal submissions.

In this way, the change in nose shape with the nostrils pointing downward may well have occurred after the control of fire, or the other way around. Additionally, some of these events may have occurred simultaneously with the increase in brain size. The order in the list aims to provide an overview of how I see the events that I think have occurred in that period. It's as if I wanted to show an outline of a house plan before showing my finished drawing. Later in the book, I will explain the process according to my ideas, elucidating more details about the chronology of events, including explainations for each event and doubts about its placement in the evolutionary timeline. For now, an overview and some absolutely essential comments are enough.

73

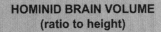

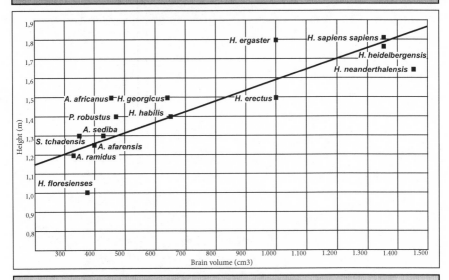

Figure 7

Since I did not agree that those small advances in manual skills were significant steps toward human intelligence, I began to wonder what had really happened. It was only when I developed the first rationalizations of the Human Intelligence Emergence Theory, based on the Raffle Theory, that I began to admit that things might not have occurred as they appeared. So, I came to have another view of the hominids. Was the brain growing? It was. The rest of the body was growing, too. Although science considers brain growth to have been unusual, it had grown along with the rest of the body, as I show in the graph of Figure 7, which I have outlined with data from various publications and where I relate the increase in brain volume and increase in the height of hominids.

Were the tools really increasing in sophistication along with brain growth? Most authors, while stressing that the sophistication was strangely small, agree. For me, these changes were too small to be attributed to brain growth Or to explain the increase in brain volume according to reverse reasoning.

Nail monkey (*Sapajus libidinosus*) - Primate found in Brazil (Northeast and parts of the states of Bahia, Minas Gerais, Tocantins and Goiás). Some groups found in Serra da Capivara, Piauí use stones to chip at other stones, which they then use as tools to crush stalks and hard fruits, and to prepare sticks to hunt larvae and geckos and remove honey. Photos licensed by Jéferson Elias Viveiros Pinheiro.

Figure 8

I think we went through the whole trajectory a little more than 2.4 million years ago doing what one would expect from the only beings on the planet that had absolutely free hands: giving them usefulness. The only exception is perhaps the kangaroo, which is restricted to Australia. It feeds primarily on fruits and vegetables and has very short forelegs.

Moreover, if we compare with some animals, we will find that more sophisticated cognition was not required for the making of those tools. I even think that the increase in the complexity of hominids' stone tools was very small in relation to the activities of other animals. Beavers build impressive dams; birds are architects that build complex nests to attract females; chimpanzees use rocks to crack open nut-like fruits; the crows of New Caledonia, living in an archipelago in the South Pacific Ocean, make and use tools (sticks) to catch beetle larvae; the nail monkeys (*Sapajus libidinosus*), Figure 8, in Piauí (Brazil), with specially chosen stones, break coconuts. The "nail monkeys" (macacos-prego) is the name given to several primates of the *Sapajus* and *Cebus* genera that are found in Brazil, and resemble the "capuchin monkeys" (*Cebus capucinus*), who are primarily found in North America.

Also, there are several birds that build extraordinarily elaborate nests. Thus, I am convinced that brain growth has no direct relation to the emergence of human intelligence, but for the time being I will not present my proposal to explain the phenomenon of extraordinary brain growth, which I wait to do in Part III where I describe other characteristics in further detail.

With the remarkable growth of the brain not being the cause of the emergence of the complex cognition of humans, some other phenomenon must have occurred. After all, in just over 3.5 billion years of the evolution of living things, there had never been an increase in cognition like the one that happened to humans. Dinosaurs, for example, dominated the planet for about 160 million years and left nothing behind to indicate a degree of intelligence similar to that of modern humans. With these reflections, I came to think of a cause that would satisfy an acceptable logic for the emergence of exceptional human intelligence.

7

My proposal

Departing from the most accepted scientific explanation for the emergence of human intelligence, I then began to seek a solution that explained how it developed, without necessarily having started with bipedal walking 4.5 million years ago or with the onset of extraordinary brain growth 2.5 million years ago. When it comes to dates, I once again emphasize that one must consider the possibility of errors in dating and that these errors can lead to incorrect reasoning. For example, some years ago, it was estimated that bipedalism had emerged around 2 million years ago. New fossil evidence, however, are pushing the beginning of this locomotive stance back to around 6 million years ago, practically when our ancestors separated from chimpanzees's ancestors. Therefore, I consider these dates approximations since they are fruits of evaluations not always reliable and often even challenged by other currents of scientific thought. I have therefore chosen to use the most accepted dates, or an average of the most accepted ones, and to only comment on differences when I consider them important for the development of my explanations.

I believe that by believing in the Raffle Theory, which I will elaborate on later in this book (chapters 28 and 29), I opened my mind to a different view of the cognitive processes of animals, including humans. This is how I came to make comparisons between the intelligences of animals. I became convinced that the intelligences of most animals are very similar to one another but extraordinarily different from that of humans. That is, humans are exceptionally smarter than all other animals, and other animals have intelligences that are very similar to one another. This puts the so-called great apes (chimpanzees, bonobos, gorillas, and orangutans) on the same level as other so-called higher animals, which science certainly does not accept today because it runs counter to the widely held idea that the great apes are smarter than other animals, especially because they are our closest cousins.

I think that primates's resemblance to us gives the impression that everything they do that is marginally different from the usual originate from an intelligence superior to the intelligence of other animals. In short, because they look like modern humans, everything they do seems like something modern humans would do and, of course, seems something clever since modern humans are intelligent. This, in my opinion, is distorting the correct observation, which is that primates, especially the great apes (with the exception of except modern humans), have intelligences similar to those of other so-called greater or superior animals such as such as cats, dolphins, dogs, wolves, lions, jaguars, and elephants.

To be clear, I am not claiming that they possess equal intelligence. I say that they have similar intelligences. However, a dolphin may have a higher intelligence than that of a tortoise, or a jaguar may have a higher intelligence than that of a snake. I may even consider that chimpanzees, and indeed other non-human primates, may have a somewhat superior intelligence to those of other animals; however, nothing indicates a correlation of primate intelligence with human intelligence. To consider primates to have an intelligence superior to the intelligences of other so-called higher animals, it is somehow to be understood that, before we separated from the monkeys of the new world 40 million years ago, we were already showing signs of superior intelligence or even the beginning of a superior intelligence. To me, this has neither sense nor logic. It points to a kind of intentionality that does not exist in the natural selection of Darwin and Wallace. Even if we consider only the so-called great primates, I would point to a direction that, unstoppable in my view, would have begun with our separation from the orangutans 14 million years ago.

I always read articles on chimpanzee research that compares their behavior with human behavior. A few years ago, I learned through the Internet and television about an experiment with a chimpanzee that could beat humans in a game that involved memorizing the positions of objects on a computer screen. The fact that a chimp can play on a computer is already impressive, I admit. But I also recognize that simply memorizing some objects (or numbers, which from the chimpanzee's point of view, I think, is the same) is not so important. It is easy to presume that an arboreal animal like the chimpanzee, which feeds basically on fruits and requires a different photographic vision, can overcome humans, who, for thousands of years, have domesticated animals and cultivated agriculture to feed themselves, without the urgent need to seek or collect almost any fruit. Even if it is impressive for a human to see a chimpanzee play on a computer, it must be borne in mind that in months or years of training it is possible to get animals to do extraordinary things, as happens - or did - in the best circuses in the world.

Nevertheless, as I was writing the first few pages of this book, I watched a documentary, *Genius Animals - Pigs* (Nat Geo WILD Channel), which reinforces my ideas about the similarities of the intelligence of various animals. The video argued that pigs are very intelligent, perhaps even more so than chimpanzees. In a computer game, where the pig participates with a joystick moved by it snout or mouth, it does better than most of the chimpanzees tested, and the only dog that participated was only able to play guided by its trainer. Several pigs are shown obeying orders and performing tricks. I found the "engendered" solution, of a trained a pig to put a golf ball inside a plastic ring, to be very impressive. Since it was not able to carry out the mission, because it's really difficult for a pig to guide a ball with its muzzle, it solved the problem differently: with its mouth, placing the hoop over the ball. Even considering these very interesting and valid experiments, I continue to think that they are only trained animals that have no intelligence resembling that of modern humans. I want to say a word in defense of the one dog that participated in the experiment. I think it needed its trainer, not because it had a lower cognition than chipmunks and pigs, but because dogs, domesticated a few thousand years ago, are very obedient to humans. It needed the coach's encouragement for each stage of the mission. The coach, therefore, wasn't teaching the dog; he was authorizing it.

I know that it's very difficult to measure the intelligence of animals, but if it was possible, I imagine that there would be small differences among them until reaching humans, when such a superior intelligence would be detected that it would be easy to distinguish it from other animals. Does any animal write; draw; paint; compose music; write sheet music; or construct buildings, ships, and spaceships like humans do? The answer is no. Human intelligence is vastly superior to that of other animals. It was this superiority that led me to start looking for what really was different in comparing human intelligence with that of other animals.

In my house, I raise dogs. I use the term "dog" because I don't like the term "hound," which in Brazilian Portuguese also denotes the Devil, the greatest enemy of God. Dogs don't deserve this. My main objective in raising them is security, because in Brazil, as in most countries in the world, it is unsafe for humans to live among humans. The dogs are my house guards. They protect me and my children not from other animals, as would seem logical 45,000 years ago, but from other humans who would not hesitate to kill in order to steal something. However, perhaps because they have been living with humans for thousands of years, dogs end up earning a great deal of affection, and, certainly because of that, I have much affection for them. My favorite dogs are the mix of Brazilian Mastiffs and Rottweilers. They are obedient, affectionate, and docile toward their owners, but very aggressive toward strangers.

Beginning when they are puppies, I teach them to enter their kennel, a structure with bars away from the house, every time I am visited by people strangers to them. This learning process, repeated daily, takes no more than 10 days. When someone visits, I go to their door and I call them by their names: "Ártico, Pantera, and Jumenta" (the first generation) or "Bite, Ártica, Lambido, Lambona, and Desmilinguida" (the second generation) and I say to them, "Inside, inside." After they learn, they invariably obey.

It was by watching my dogs enter their kennel that I tried to differentiate the actions dogs can perform from the actions humans can. I came to the conclusion, using computer analogy, that two major differences separate human intelligence from that of other animals: the space available for recording information and the software to work with that information and provide solutions. Because of this, I began my second analogy of the cognition of living beings with computers. The first was when I elaborated the Raffle Theory.

I will use the term "algorithm" instead of "software" because, although the definitions are similar, algorithm gives the idea of a set of instructions still in the development phase of software, an idea that, in my view, is more compatible with the natural selection process of Darwin and Wallace. It can be said that software is composed of a set of algorithms and that an algorithm is composed of a series of instructions. In this light, software can be seen as a great, diversified algorithm. But it was the notion that software is something finalized while an algorithm is something still in development that caused me to opt for algorithm in the analogy with the evolution of human cognition. The reader should note that the term algorithm is used here in the sense of a set of instructions capable of working with a certain amount of data in order to define and present solutions to problems. It is something very similar to a computer program routine. A computer program routine can, for example, provide the salary of any employee of a company as long as one or more pieces of information that correctly specifies such salary is correctly provided. But, for this, it needs two essential things: space to store all of the names of the company's employees with their respective salaries, and an algorithm with instructions capable of searching for an employee's name and providing his or her salary. The explanation is simple enough, but it fits perfectly into the reasoning. Dogs cannot obey orders to enter 10 out of 100 rooms available each day in a group of many rooms because their brains do not have adequate space to store so much information. Nor do they have the algorithms capable of working with so much data to solve this problem. Therefore, they can only perform the simplest tasks, which involve little data. Humans, however, would carry out these orders easily, entering different indicated rooms every day, without any problem.

Humans understand thousands of words because they have room to store them, and dogs can learn only a few words because they only have space to store a few words. Similarly, humans have algorithms to work with thousands of words (or pieces of information), and dogs have algorithms that can only work with a few. This corresponds to the reality presented by research: the average vocabulary used by modern humans today is a thousand words. Dictionaries contain around 142 thousand entries, while for the most intensely studied mammal, the green monkey, only 10 calls have been distinguished (DIAMOND, Jared. *The Third Chimpanzee*, p. 168).

In any case, I had a point of departure. Now, I needed to find out when, how, and why human brains acquired the space to store large amounts of information, creating conditions for algorithms to work with information, and to return acceptable answers to complex problems that could record thousands of words, understand graphs, and perform calculations. Today, already accustomed to the reasoning that led me to the Human Intelligence Emergence Theory, it seems all logical, natural, and intelligible. However, when the idea first came to me, I confess that I considered it very difficult to fit within the random process of natural selection.

I will detail as clearly as possible the computer analogy and begin by explaining why I believe that this kind of analogy makes sense. Humans better understand things that can be compared to things they produce. When someone says that a cheetah runs at 120 km per hour, the first comparison we make is with an automobile, which is the human product that travels at a similar speed, in order to understand how fast the cheetah runs. In this case, the cheetah's muscles and feet compare to the engine and wheels of the car. Reverse reasoning also helps. Saying that a certain crane does the work of 20 men helps in understanding the problem for both those who understand the crane and those who understand human work. It is not very difficult to understand why some scientists compare the human brain to the computer: they both solve very similar problems. The brain and the computer make calculations, report an employee's salaries, draw graphs, draw floor plans, and produce information. They have many points in common for a comparison and, logically, for an analogy. And, when we compare, as much as we may feel offended, computers come out better in various ways. But this isn't a subject for this book.

Let us continue with the analogy. Since the example I will use now is a computer, I return temporarily to use the term "software." Imagine a person, with a computer but without spreadsheet software and with a 120 GB hard disk (HD), needing to draw up a payroll. He goes to a computer store and is told that spreadsheet software with all payroll data needs 130 GB of HD in order to work. There is no point in buying only the software, since 120GB of HD will not hold the payroll data. It is also useless to buy only a larger HD, say 150GB, because the computer without the software will be unable to process the data. The only way to solve the problem would be to buy both the software and an HD that would fit the data. Then, the employee makes the right decision and, finally, creates his spreadsheet.

In nature, things do not happen this way because there is nothing analogous to an employee to direct the two decisions at the same time: the acquisition of software and the increase of space for recording data. In nature, everything happens at random. In addition, energy savings are very important, and changes that consume a lot of energy, if they don't have great relevance for the evolutionary moment, end up discarded.

Incidentally, the HD of the human brain (let us call the data space in the brain this way to facilitate the reasoning) could increase a little in size, but it would need, also incidentally, software in the brain to take advantage of that space, or else the increase would be discarded in the next generations for the lack of utility. When I speak in terms of HD of the human brain, it is clear that I am referring to the system that the human brain uses to store data, which in no way needs to resemble the system that a computer hard drive uses to store data. The comparison is restricted to the end, not the method. The same reasoning applies to the use of the terms "software" and "algorithm," which here refer exclusively to the set of instructions capable of working with data and producing logical results.

Also by chance, software could also appear that would process a language like that of humans, but if it did not happen to appear soon enough, with enough space for it to work and to use minimally necessary vocabulary, it would be discarded in future generations due to lack of use. I repeat the term "incidentally" so that it is as clear as possible that the occurrence of these random phenomena at the same time would be very difficult to succeed in nature, perhaps impossible. It also appears that it was so. Dinosaurs dominated the world for 135 million years, but they did not acquire intelligence similar to that of modern humans because there was never an immediate advantage for their brains to emit large amounts of data while immediate evolutionary advantages would provoke the emergence of software capable of taking advantage of spaces to save and recover new data.

Mammals have been around for millions of years, and only one of them developed intelligence similar to that of modern humans: modern humans themselves, precisely because of the improbability of the appropriate conditions for the emergence of a human-like intelligence. My problem was figuring out how and when these conditions occurred.

When I speak of the emergence of spaces for recording large amounts of data, I'm not necessarily talking about creating new physical spaces for recording data or a redistribution of memory spaces. The space for recording data may have increased without the increase of a single neuron (exaggeration justified by the explanation): simply with the emergence of an algorithm able to control the use of memory in the brain for calculations, data storage, or an algorithm capable of compressing data in an already existing memory space.

The best way to understand is with a simple example: the American Jill Price, at age 12, noticed that she remembered everything that had happened in her life, in detail. Everything! Day by day, she remembered all the news in the newspapers, whether the events were important or not. Anguished by the fact that memory was taking over her life, she sought help and found, through Google, scientist James Gaugh of the University of California. Gaugh and his team found that twenty areas of Jill's brain are larger on average than those of most humans. They also found an anomalous lateralization, a trait of autism—a disease that she does not have—that leads to a mix in the division of tasks between the right and left hemispheres of the brain. That is to say the following, I imagine: Price's brain had not undergone any extraordinary physical modification, but it managed to store and retrieve a quantity of relational and related data unimaginable to another human being. What exactly happened? Gaugh's team found no answer. Neither do I, nor most likely the reader, nor the scientists, from what I see on the news. It is still a mystery!

So, when I speak of increasing memory spaces for data recording, I am talking about increasing the data space available for the operating system that commands the human brain. How this data is made available to the system has little to do with it. What matters is that they are made available where they were not before.

It was when I tried to understand how one could move from intelligence like that of a dog, a lion, or a chimpanzee to an intelligence like that of modern humans that I watched the DVD *Evolution - the Adventure of Life* (Production of Miles Barton, BBC, 2005).

In film 5 (Human Beings) appeared a fact that intrigued me. I transcribe the following caption from 17 to 20 minutes:

"As we began to live in groups for safety, our brains grew to deal with it. And the brains kept growing until they increased four times in volume. So our skulls had to grow too. This left us with the next challenge. After all, being the savage biped "intellectuals" created a problem for us that our quadruped friends didn't have. Take wildebeests, for example. Since they walk on all fours, they can support wide hips. This makes birth quick and easy, which is important when you have so many predators around you. Although the wildebeest baby is large, its brain and skull are quite small in relation to the size of its body. Therefore, it easily slides out of the mother. But, for us, it is during birth that we pay the price for having big brains. Moving from the savannas of Kenya to the city of Nairobi and observing the silhouette of modern women, we can understand why. The ideal build for standing upright is a slender body with narrow hips. However, narrow hips are the worst possible option at the time of giving birth to humans with an advanced brain. **This was one of the biggest obstacles we encountered throughout the adventure of life. We were stuck. Until evolution came up with a surprising solution. Unlike nearly all other mammals, our skull is not fully formed when we are born. The human baby's head is soft and malleable, which allows it to pass through the narrow birth canal. Throughout the first year of life, the bony plaques of the skull gradually join together to protect the brain. In fact, we are all born premature.** So, in the early months, the human baby is extremely vulnerable and requires constant care. But the advantage this entails, a giant brain, far outweighs the risks. Yet another biological dilemma was solved. So, 300,000 years ago, Homo sapiens and their large brain appeared."

In the caption's bold text, the scenes of a human birth, and an animated illustration are shown interspersedly, illustrating the difficulty of the birth of humans and how much the baby's brain is malleable in order to be able to endure pressure at the moment of birth, most of the time, without damaging its structure. The statement **"Actually, we're all born premature"** is quite startling, at least to me. In Figure 9, I present a summary of the sequence of images of the animation that inspired me to form the first concepts of Human Intelligence Emergence Theory and to pay homage to everyone who participated in the roduction of the documentary *Evolution - the Adventure of Life*.

A few days later, my brain returned an amazing solution. Let the reader observe the preceding sentence and see that I have not resisted my habit of

Animation image sequence
from the documentary Evolution - the Adventure of Life

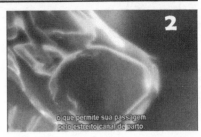

1 and 2 - the baby's head approaches the vaginal opening ...

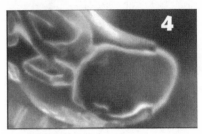

3 and 4 - malleable, undergoes a kneading, in order to pass through the narrow human pelvis ...

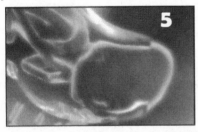
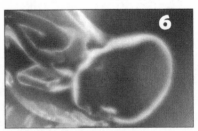

5 and 6 - face down, in a difficult move, to allow the shoulders to pass ...

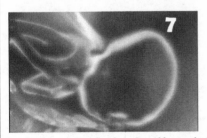

7 and 8 - overcomes the obstacle and leaves the body of the mother, returning to the normal size.

Figure 9

which I spoke in the Introduction when I referred to my brain in the third person, as I have habitually done in recent years since I came to believe in the Raffle Theory. The explanation of these facts could be different. Instead of brain growth causing hominids to be born prematurely so that the NEW big brain could, at birth, pass through the small pelvis, as the explanation of the DVD says and is the general view of science, events could have occurred in a different way, in two important and distinct stages. First, probably starting 2.5 million years ago, the brain would have grown and brought about a proportional enlargement of the pelvis for the easy birth of babies. Only, after this, an urgent narrowing of the pelvis would have made the hominids be born prematurely, so that the brain, which was already large, could pass through the NEW, narrow pelvis.

I will explain further. Imagine that our ancestors, with their large brains, and, of course, their large pelvis proportional to the size of the brain, due to some climatic phenomenon, needed to survive and run and/or swim very efficiently, and that for this reason, there was an extreme need for the pelvic reduction in order to allow a leg movement that would make both swimming and running more efficient. The most logical solution would be to narrow the pelvis, to facilitate the urgency of running and/or swimming, and to reduce the size of the brain to make birth possible in the face of the new situation. But it seems that brain size is very important to hominids, and so evolution would have to look for another way. The path chosen, according to my theory, would have been that of **prematurity**, which caused our babies to be born prematurely, with smaller and more malleable brains and the cranial box not yet completely closed, thus allowing passage through the birth canal, and then, after birth, to complete the protection of the skull with the junction of the cranial plaques. This would complete the stages of development of the brain, which were previously made when the baby was still in the womb.

It turns out that evolution does not seek a single path. In fact, it searches for several paths, and usually only the main ones that caused the phenomenon we are studying are retained. I believe that evolution, in order to solve the problem, has mainly used two paths to respond to the urgency of narrowing the pelvis: the first one, as I have already said, the most logical, was to reduce the size of the brain so that the delivery would occur without problems; and the second, which seems more ingenious to me, was to birth humans prematurely so that they would still be born with the small, malleable brain to get through the new, narrower pelvis. But the importance of brain size for hominids/humans slowed down the first option and consequently accelerated the second. I used the expression hominid/human because I consider that only here, according to my theory, were hominids becoming human as we know them today, or modern humans, as many authors prefer and as I have been calling them in this book.

When I say that evolution has slowed down the first option, of lowering the size of the brain to pass through the new, small pelvis, I do not mean that it has not been used. On the contrary, since it was more logical, it must have been the first one to be taken and replaced only when the option of prematurity proved more effective and less harmful, especially since it had the advantage of maintaining brain size. **Thus, if I am correct, the emergence of human intelligence, which I think really began at this point in evolution, would be paradoxically associated with a small decrease in brain size, not an increase, as is widely accepted.** There must have been a hominid on the African continent, an ancestor of modern humans, with the brain at least slightly larger than the brain of *Homo sapiens*.

I already knew that *Homo neanderthalensis* had a brain more than slightly larger than that of *Homo sapiens*, based on a skull found in Amud, an Israeli site, with the extraordinary volume of 1,750 cm3 (ARSUAGA, Juan Luis. *Neanderthal's Necklace*, p. 96). But this skull must be an exception. In any case, after calculating an arithmetic mean using data from various publications, even imagining that surely, and perhaps unconsciously, there is a natural tendency to maintain human supremacy in relation to brain size, I came up with a result that shows the *Neanderthals* to be with an advantage. Modern humans have an average of about 1,400 cm3, and *Neanderthals* with 1,520 cm3.

But the size of the *Neanderthals*' brains being larger than those of modern humans does not help prove my theory. It is widely accepted by science that the ancestors of *Homo neanderthalensis* separated themselves from the ancestors of *Homo sapiens* about 500,000 years ago when the ancestors of the former migrated from Africa to Europe and were only once again found around 40,000 years ago when *Homo sapiens* arrived in Europe. So, even though the *Neanderthals* did mate with the *Sapiens* when they met in Europe,

Three angles of a replica *Homo sapiens idaltu* skull, acquired from Bone Clones, Inc. Photos licensed by Jéferson Elias Viveiros.

Figure 10

as some scientists believe, this would have no bearing on my prediction, because *Neanderthals* would be the ancestors of modern humans, not of *Homo sapiens*. **According to this understanding, I should look for some fossil evidence that there was an ancestral** *Homo sapiens* **hominid with a brain larger than that of modern humans, most likely on the African continent.**

I began to search the Internet for a fossil that supported my prediction, unfortunately to no avail. I asked the search engines the most varied questions, but the best I could get was the information that fossils between 500,000 and 40,000 years ago, the range of *Neanderthal* separation and the emergence of the first archeological records of human intelligence, were rare. My inexperience about the nomenclature of hominids blinded me. I lacked a regular undergraduate course treating human evolution. Until, by doing my research on the Internet, I came upon an article in the Brazilian weekly magazine Veja online titled *Adam was African*. One sentence from this article struck me, especially since it referred to fossils dated 160,000 years ago from three well-preserved skulls of two adults and one child found in 1997 in a hamlet called Herto in the Afar region of Africa (Ethiopia): "Their skulls were a little larger, the brain slightly larger and the face longer than those of the modern man." Discovered by Tim White and colleagues, they were classified as *Homo sapiens idaltu*, Figure 10, and that was when I understood why it took me so long to find it.

I had been mistakenly looking for a hominid before the *Sapiens*, but I didn't know that it had been already classified as *Sapiens*. In fact, I admit, I was not very smart in this search, for I should have imagined that the hominid I was looking for should have been classified as *Sapiens*, because, after 200,000 years ago, almost all of our ancestors are classified as *Sapiens*. Moreover, the only differences I predicted were two small variations: one in brain size, and another in the opening of the pelvis, probably insufficient to be considered a new species.

With the correct name, it was easy for me to look for several texts about him, which confirmed the information published in Veja online. If this evidence did not corroborate my thoughts, at least it was a fact that pointed to something I had anticipated, perhaps an indication that my theory might at least deserve to be better examined. **After all, an archaic man dating to 160,000 years ago did have a slightly larger brain than did modern humans, as my theory had assumed.** However, I was not too optimistic. Those hominids could be an exceptional group, people with some brain disease or even *Neanderthals* with some anomaly that confused them with sapiens, who eventually would have arrived in Africa. Only new fossils could confirm my ideas.

In fact, there is strong evidence that somehow, we had an important adaptation related to the large brain and the narrow pelvis between 500,000 and 40,000 years ago. In addition to the time of gestation, other factors were altered in the procreation of humans, which made our birth very different from the birth of chimpanzees, our closest living relatives (DIAMOND, Jared. *The Third Chimpanzee*, pp. 92 and 93). In females, for example, when they approach mating age, a process of hip enlargement occurs, probably to facilitate childbirth. Males' shoulders broaden, perhaps to facilitate swimming and running, important for aquatic and terrestrial hunting, respectively, to provide offspring with the necessary proteins.

Humans and chimpanzees (I have no information about other the great primates), unlike most monkeys, are born with their faces turned upside down. However, because chimpanzees have a proportionately larger pelvis than babies' brains, their females perform a quiet, solitary birth. Childbirth in humans, however, is difficult and painful, which makes the females of virtually all known cultures seek help in giving birth to their offspring. The birth canal, the bony opening of the pelvis through which the human baby passes during birth, is almost the size of the brain and has an oval and irregular shape in its extension, causing the human baby to perform a series of manipulations, including a 90° turn to pass the larger parts of its body—the head and shoulders that must always be aligned with the widest part of the canal throughout its extension. I think this reinforces my idea that it was the pelvis that changed due to extraordinary evolutionary pressure, requiring changes in the human birth process, prematurity being one of them.

Explaining in further detail, I propose a change in the order of events established that brain growth caused prematurity in order to solve the problem of the passage of a new, large brain through a small pelvis. I propose **a completely different order, divided into two periods: first, the brain grew, causing an increase in the opening of the pelvis (between 2.5 million and 120 thousand years ago, approximately); then, an urgent decrease in pelvic opening caused prematurity, which solved the birthing problem of a large brain through a new and small pelvis (approximately between 120 thousand and 40 thousand years ago).**

I take that position precisely because, according to the solution my brain has reached, therein lies the key to the explanation of the emergence of human intelligence, which I shall explain shortly. May the reader pay attention to the first explanation: the brain needed to grow in order to provide humans with intelligence. This argues that, for nature, human intelligence was absolutely necessary, and that the small pelvis was impeding it because it hindered the growth of the brain, and, to solve the problem, we began to be born prematurely. Brain growth would then be an absolute requirement to provide humans with intelligence, and prematurity would be a necessary consequence to allow the birth of babies, who were born with increasingly larger brains.

My explanation changes everything.

First, hominids' brains experienced significant growth, from 2.5 million to around 120 thousand years ago, for reasons that I will try to explain further on, and only about 50 to 40 thousand years ago did the emergence of a differentiated intelligence appear on the planet.

Second, during the same period, the opening of the pelvis expanded in conjunction with brain growth, always allowing for an easy delivery, as with most great primates.

Third: Hominids, before acquiring human intelligence, had a pelvis perfectly adjusted to give birth to babies with large brains.

Fourth, an urgent need (for running and/or swimming, for example) required an urgent reduction in the opening of the pelvis to facilitate the necessary leg movements for these types of locomotion.

Fifth: hominids came to be born prematurely, so that the smaller brain, still in stages of formation, could pass through the now narrow pelvis.

According to my line of reasoning, some archaic humans, our ancestors, had large brains—even a little larger than those of modern humans—but did not have the human intelligence of modern humans. The extraordinary increase in brain size and the concomitant small technological advance in the making of stone tools, therefore, had nothing to do with a drive toward cognition similar to that of modern humans. Here is the great rupture of my ideas with what is now established by science: **for me, the size of the brain has nothing to do with human intelligence.**

But the reader should not understand this "nothing" to be very strict. I admit to having used such a radical term more to attract attention. Of course, there cannot be an intelligent brain that does not require the occupation of some limiting space either in absolute or proportional size. What I affirm, and I acknowledge that this is already a very great extravagance because of currently accepted thinking, is that an intelligence similar to that of modern humans could arise in several animals that do not possess such a large brain in absolute size or in proportion to the rest of the body compared to humans. Thus, I also argue that a human-like intelligence could emerge in various animals such as monkeys, dolphins, dogs, chimpanzees, gorillas, goats, lions, elephants, and chickens. The size of the brain, according to my thinking, has greater importance within the same species, because the space to store information, or to process it, makes a difference when one has the same technological data processing apparatus. Between two species, there can and should be technological differences, and this can only be shown by way of example because it is opportune for reasoning of how a smaller brain is able to store more information than a larger brain.

That reminds me of pygmies. But there is an explanation for them that, at least apparently, makes them unfeasible as an example for this reflection. Pygmies have small bodies and large brains due to a slowing of growth during puberty, when the brain has already reached its definitive size (WONG, Kate. *The Smallest of Humans*). In any case, I would like to have access to a more thorough study of the brain size of pygmies of the various tribes still existing in Africa to whether there really is no decrease in brain size because of the deceleration in body growth. Perhaps a better example for the reflection I made at the end of the previous paragraph is the fossils of very small hominids: about one meter high and with 380 cm3 cranial capacity, found in 2003 on Flores Island, Indonesia, dated 12,000 years ago, dubbed *Hoobit*, and classified as a new species—*Homo floresiensis* (see table in Chapter 1).

This has been causing great controversy because some scholars think it is a new species that probably originated from *Homo erectus*, while others simply think that it is a *Homo sapien* that has undergone a process of "island dwarfing" because some tools found with the fossils are characteristic of *Homo sapiens*. Island dwarfism is an artifice of nature, not yet properly understood, that causes some beings to diminish in size due to being isolated on an island.

Then, how could such a small brain have intelligence comparable with that of modern humans, as the tools found in their caves demonstrate? What do I think? I prefer the second opinion because my theory would hardly align with the idea that *Homo erectus* survived up to 12,000 years in Indonesia. I do not think they are simply *Homo sapiens*. They are modern humans who suffered the phenomenon of island dwarfism: their brains diminished, maintaining all the new arrangement that provided the human intelligence, but in a smaller space. In short, nature miniaturized the brain of modern humans. Depending on the momentary pressure, I admit that it may have diminished cognition, but given its evolutionary advantages, I do not believe that this decrease was enough for the *Hoobit* to cease to be a modern human. In any case, my comment about *Floresiensis* is just an opinion, and certainly the debate about its classification will last for years, until perhaps some new fossil find may clarify the issue.

I repeat, intelligence similar to that of humans, according to my studies, could have arisen in a number of animals, without the slightest commitment to the emergence of bipedalism or to the enlargement of the brain and without cause and effect to other human characteristics, and in some cases, even with an advantage. For example, if the phenomenon of the appearance of human-like intelligence had occurred with the nail monkeys of the Brazilian

93

Northeast, where I live, they would have several advantages because they are much smaller and lighter than humans, requiring lesser food and occupying lesser space. They could even move around more easily, using technological advances achieved by modern humans. Given its weight, I think the personal helicopter would be a common vehicle in the skies of big cities. Imagine how many nail monkeys would fit on an airplane or ship. They could even use the computer more efficiently, using the mouse with their tails and leaving their hands free for typing.

Of course, this is just an example. It may even be that nail monkeys have some physiological impediment of which I am not aware that would prevent them from becoming intelligent like humans, other than just the size of the brain.

Having completed my reasoning on the order of events, the reader might ask, where does human intelligence come in? I will explain. **In my opinion, prematurity was not a consequence of the emergence of human intelligence. It was, in fact, the very cause! Or at least the biggest cause: the engine that fueled the process, as it were.** Our brains were not growing, making us more intelligent. We were not being born more and more prematurely so that the new, big brains could pass through the pelvis. My proposal is that the pelvis was getting narrower due to urgent ecological factors, and we were born more and more prematurely so that the brain, smaller because it was still in formation, could pass through the new, narrow pelvis. **That is, the process of the emergence of human intelligence did not cause prematurity. Prematurity caused the process of the emergence of human intelligence, as I will explain later.**

Here, I think it necessary to conceptualize instinct and culture side by side in order to facilitate understanding my ideas. Generally, the term "culture" is used in the generic sense, which includes knowledge, beliefs, the arts, morals, laws, customs, and all other habits and aptitudes acquired by man as a member of a society, which confuses and mixes the concepts of culture and instinct that some scientists adopt and that I follow. Because of this and because of the importance they have for understanding the ideas I will attempt to set forth, I will clearly define the concepts of the two terms for this book, pluralizing them in order to facilitate my explanation: instincts are all the knowledge and behaviors that living beings receive from their ancestors before birth, by genetic or other means still unknown; cultures are all the knowledge and behaviors that living beings conceive of or receive from other living beings after they are born.

A brain is born with a variety of instincts. Some appear soon after birth and others appear in the course of life. The great peculiarity of instincts in relation to cultures is that we are already born with them. The moment these instincts arise does not matter much. Some examples of instincts are the instinct to suckle and kill (humans, dogs, lions, cats, etc.), the instinct to stand up immediately after birth (wildebeest, horses, zebras, etc.), the instinct to eat right after birth (chickens, Guinea Hens, peacocks, etc.), the instinct to push out (i.e., killing) other birds in the nest, and the instinct to make nests (birds).

Clearly, these concepts serve to facilitate understanding, but, in reality, on specific occasions, instinct and culture are mixed, especially when trying to define the moment in which life arises or to understand the genetic, chemical, electrical, or other, still unknown qualities, which transfer or can transfer physical or behavioral characteristics between living beings. Sometimes, however much we try to separate them, instinct and culture intertwine, and we do not know where that information or behavior actually came from, whether it is inborn, received from another living being, or deduced by extraordinary human intelligence (RIDLEY, Matt. *Nature via Nurture*, p. 76).

With the human species, one is never certain that what one sees is instinct, because one may be looking at the result of a reasoned argument, a copied ritual, or a lesson learned, just as one can never be sure that what one is seeing is only cultural, because it may have an element of instinct (RIDLEY, Matt. *Nature via Nurture*, p. 76). Nevertheless, for my explanation, these concepts are essential and efficient—even to understand how, in many cases, they influence one another, mingle, and become confused, especially when one analyzes the behavioral outcome of humans.

Returning to the analogy of the brain with computers, by the definitions of instinct and culture, it is not difficult to understand instincts as already be engraved within the human uterus—similar to a computer's ROM memory, which at least in theory, cannot be erased—and culture as recorded in the brain's memory most similar to the computer's RAM—which can be recorded, erased, and rewritten. Generally, just after they are born, humans need only to be placed close to the mother's breast to start suckling. This action stems

95

from information recorded in the womb. Furthermore, commonly, humans are afraid of spiders and snakes, even if they have never seen any such animals during their lifetime. They can scarcely overcome this fear, even after they become adults. It is, therefore, information recorded in the womb, in the spaces of memory that cannot be erased. By this reasoning, suckling and being afraid of spiders and snakes are instincts.

However, all humans spend their days learning things and, depending on whether the new information is important for their lives, they may forget it even before the end of the day. Now, we are talking about cultures, which are recorded in erasable memory. A single example might better clarify the question: a 10-year-old male child studies regularly in school. Almost every day, the teacher teaches him to speak, write, and do math. This is culture. But his eyes, his attention, and his dreams are almost always involved with that annoying but beautiful girl who sits in the second row and doesn't take her eyes off him. This is instinct—an instinct that was engraved in that child before birth but only appears at a certain moment. To surface, one needs a condition. In this case, the condition, at least the main one, seems to be the age of the child. But it may be another. Even before he had the, let us say, physiological conditions to procreate, his instincts propelled him toward that girl. He doesn't yet know the disappointment he will face because, as she approaches puberty, she will become interested in older boys who already possess physiological conditions to impregnate her. That, too, is instinct.

So, I think it is more than logical to relate **instinct** (behavioral attributes that all humans have) to computer ROM, a virtually unchanging memory that is accessed the moment you turn the machine on, and **culture** (knowledge and behaviors that humans acquire during life) with computer RAM, a memory that is born clean and is rewritable. Note that although the terms are in the singular, they have acquired a plural meaning. Also, here, I am making an analogy. I am not comparing the physical structures of chips with neurons. Nor should one consider as an absolute rule the condition of memory being changeable or unchangeable. There are instincts that suffer such a great influence from the culture that they almost disappear, as is the case with the instinct to kill, which is lost in some people because of certain religious cultures. There are cultures that are established in such a way in some groups of people that they are confused with instincts.

Understanding culture and instinct as gradual characteristics, not as fixed properties, despite the play on words, certainly helps to assimilate my positioning. To understand the ideas in this book, it is more than enough that the reader imagines that from conception on, we begin to have some memories less and some more susceptible to change and even blank memories for recording new data.

The reasoning that made my brain conclude this was that it is quite logical that the vast majority of animal instincts should be recorded in the final days of gestation when the brain structure is most near to completion. It is unreasonable to think that instincts are recorded before the formation of the respective memory spaces. Therefore, it is logical that at least the vast majority of instincts should be recorded in the final days of gestation. A rather rough but didactic comparison to facilitate understanding is the school notebook and the annotation of a lesson: first you have to have the notebook and then take notes in class. **In this way, as we have tried to demonstrate previously, an extraordinary evolutionary pressure to narrow the pelvis could, in a reasonable time, cause prematurity in hominids, causing them to be born without their instincts completely engraved; thus, they would have large, blank memory spaces. We would thus have the external motor necessary to start and, more importantly, to feed an explosive process in the direction of something like human intelligence because those blank memories could now be engraved while the hominids are outside the mothers, while in contact with the environment, with all the senses working, and, above all, with all this influencing the use of the blank spaces of these memories. Prematurity, then, would be the engine that drove the rise of human intelligence, the engine that created more and more free memories, providing a lasting environment favorable for both the existing algorithms to take advantage of these new memory spaces and for the emergence of new algorithms that, perhaps, required an environment with capacity to store more and more data. It would be the enigma of human evolution—that which made us different from all living beings.**

Sometime between 500,000 and 40,000 years ago, when there was an urgent and extreme obligation to narrow the pelvis, humans should have had a simple language such as, for example, that which chimpanzees have today, with a few sounds identifying a few situations and objects. With the need to make babies' heads pass through the newly narrowed pelvis, humans came to be born prematurely; prematurity caused humans to be born with blank memories, and more free memory spaces provided the chance for language to grow its dictionary and become more sophisticated, powerful, and complex. With the urgency of the narrowing of the pelvis, humans became more and more premature, being born with more and more free memories for the emergence of new algorithms and the improvement of existing ones. Language is only an example. Comparison, imitation, counting, the differentiation of sounds, recognition of group members, recognition of other animals, multiplication, division, addition, and proportionality are examples of algorithms that work limitedly with few free memories but that become very powerful with large memory spaces.

Thus, it would only be necessary for the brain to already have an algorithm capable of using more memory spaces, or for a new algorithm to emerge, now in a new environment created by the appearance of memory spaces. Note the difference that prematurity causes to the process. Without it, two near-simultaneous events would be required to occur many times: the increase of memory spaces and the emergence of an algorithm capable of working with more memory spaces. With it, an algorithm capable of using more memory spaces would be enough because these memory spaces would always increase as a function of prematurity. Perhaps not even this, if the brain already had some algorithm with the capacity to use more memory spaces and complex cognition was restricted due to the lack of a place to store more data. This might be the case with most of the algorithms I have cited.

Again, I turn to computers. A computer that has more free memories than another can certainly house a much larger dictionary. A computer that has more free memories than others can certainly scan and organize many more images (analogous to the recognition of other animals). A computer with more free memories than another computer will certainly compute much more complex sums, multiplications, and divisions. With the engine of prematurity at full steam, producing ever more free memories, I think, something like the inflationary spiral that Richard Dawkins talks about can occur: and at an extraordinary speed, as I shall try to show in the following chapters.

This argument, if correct, suggests that hominids/humans lost several instincts in the process, because prematurity would not leave enough time for the recording of the last instincts still in the mother's womb, and the space would be harnessed by algorithms recorded after birth. It seems this is what happened: hominids/humans lost most of their instincts about independence at birth: they do not know how to hold onto their mother or even to crawl on the ground. They are highly vulnerable. They only gain some independence around eight months after birth, when they begin to take their first steps.

Now that I have laid the foundations of my theory, a very important question arises: when, more precisely, do I imagine the process was initiated, and when was it completed?

Around 500,000 years ago, according to fossils and the most accepted scientific thinking, the following hominids existed on the planet: *Homo heidelbergensis, Homo rhodesienses*, and *Homo erectus. Heidelbergensis* and *Rhodesienses* lived only in Africa. *Erectus*, the first hominid to come out of Africa, lived both in Africa, where it is called *Homo ergaster* (main fossil known as the Turkana Boy) as well as in Asia, more precisely in Georgia, China (Beijing), and Indonesia (Isle of Java). But it is probable that there were other hominids whose fossils have not yet been found.

Even accepting that other hominids lived at this time, it is certain that from this select group, only one emerged as what we call Humanity. At least, the vast majority of the scientists I have reviewed advocate this. Although there is the very remote possibilities, I may say, of humankind descending from two or from all of them. To believe in these possibilities is to accept the multiregional theory, mentioned in Chapter 1, which says that modern humans independently evolved from Homo erectus in various parts of the world. As I mentioned earlier in Chapter 1, this theory is not in line with my ideas, but follows the thinking of the overwhelming majority of scientists in defending a unique origin somewhere in Africa.

It seems to be fully accepted that our ancestors separated from Neanderthals' ancestors about 500,000 years ago, when the ancestral Heidelbergensis ascended to Europe, and our ancestral Heidelbergensis remained in Africa. Modern Neanderthals and humans only encountered each other around 40,000 years ago, when modern humans, who were already spreading around the planet some 15,000 years ago, arrived in Europe, almost certainly via the Middle East. They lived around 10,000 years ago, when the Neanderthals were already extinct.

It is also generally accepted that there were no crossings between the ancestral line of the Neanderthals and that of modern humans until 40,000 years ago, when they met again. Even admitting that there may have been some contact, given their proximity in Europe, they certainly did not leave descendants, according to genetic studies that point out that the common ancestor lived around 500,000 years ago.

However, it is good to note that some geneticists have begun to publish studies indicating that there was some cross-breeding in this 10,000-year coexistence period, but current studies do not agree with this, arguing that the DNA found in common—between 2% and 4%—came from a common ancestor and not through hybridization or cross-reproduction between two species. Nevertheless, if there were crosses, this doesn't conflict with my theory. My line of reasoning indicates that it would have been very difficult for this to have happened. First, some studies point out that the time of gestation of the Neanderthals was 12 months or a little more, which would prevent, at least in theory, the birth of a hybrid baby of modern humans, prepared to leave the uterus in 9 months. The modern human female, who already had problems with birthing humans, almost certainly could not have delivered a *Neanderthal*-hybrid child, which would certainly have had a skull larger than her pelvic opening. There must have been several other impediments because of the timespan of their separation, the *Neanderthals* in Europe and the *Sapiens* in Africa.

Thus, it is logical to think that the ancestor of the two, probably *Homo heidelbergensis*, also did not possess human intelligence. If the process of the emergence of human intelligence had already begun before *Heidelbergensis*, one would have to admit that only the lineage that resulted in modern humans had attained the complex cognition we have, giving it advantages that provided the conditions that eliminated other lineages before they developed a complex intelligence such as that of humans. As I have already pointed out, I don't see anything that leads in this direction, and here, once again, the great discrepancy of my thoughts with what is accepted today is evident. For me, there is nothing to indicate the appearance of complex cognition in hominids between 2.5 million years ago to 500 thousand years ago. So much so, the

Neanderthals, originating from the same *Heidelbergensis* from which we probably originated, never developed intelligence like ours. Of course, there is a loophole for the defense of the thesis that a complex cognition was already outlined 500,000 years ago: that the *Neanderthals* did not come to it only because they became extinct. I cannot dispute this with logical reasoning, and with my commitment to sincerity, I will continue trying to justify my line of thinking.

Somehow, the definitive foundations for the realization of modern human cognition have only left undisputed marks after a few hundred years of separation from *Neanderthals*. Logic firmly points to this, even leaving a small window for the possibility that some other extinct hominid may have developed human intelligence. The conclusion I reached was that we were similar to most of the animals that lived on the planet and perhaps even had a level of cognition lower than some of them.

To summarize, I deduced that hominids of that time had nothing exceptional to direct them towards evolving intelligence similar to that of modern humans. As indeed happens with the great majority of animals, they only had their peculiarities. Among these were bipedal walking and the exaggerated increase in brain size. Did we make tools? Yes, but today it has been proved that many other animals also make tools, even though they do not have intelligence similar to ours.

That said, I consider, with a good degree of reliability, that modern cognition did not begin to develop before 500,000 years ago. Now, I must limit what I can say with certainty, that at least one human being with an intelligence similar to that of modern humans already existed on the planet. I imagine that here, we have to believe in proven traces that have been left. When I say this, I include the most guaranteed dates. With this degree of reliability, clear marks of a human intelligence have been detected in archaeological records that are only 40,000 years old. These include spearheads, drawings, paintings, sculptures, statuettes, and adornments. Findings that point to previous dates have always left doubts, and I prefer to wait for more conclusive and guaranteed evidence before proposing an earlier date for the emergence of human intelligence. I think it really is impossible to imagine that a creature devoid of intelligence at least similar to that of modern humans could have produced, for example, the drawings of animals in the caves of Chauvet in France (35,000 years ago), the ivory carvings of an antelope and a horse found in Germany (32,000 years ago), the animal drawings of the French cave of Lascaux (17,000 years ago), and the drawings of the Serra da Capivara National Park in São Raimundo Nonato, in my Brazilian state, Piauí (currently undergoing the dating due to the great variety of findings). In any case, leaving aside this healthy struggle to find older vestiges when I elaborated the theory, I came to the span of time that I considered as certain for the emergence of human intelligence: between 500,000 and 40,000 years ago . The Great Leap Forward, an expression coined by Jared Diamond, must have occurred within this period, and I think that is exactly what I call The Rise of Human Intelligence, or as most authors prefer, The Rise of Humanity. The very creator of the phrase makes the hook in his book, The Third Chimpanzee:

"If there was a particular point in time when we can say that we became human, it was in that leap."

(DIAMOND, Jared, *The Third Chimpanzee: The Evolution and Future of the Human Animal*)

In deciding to publish my theory, I thought of simply explaining it, and, if it were correct, someone with more preparation and competence could establish the timeframe in which the process occurred. I was not going to become involved with dates. However, I continued to study more books and articles, learning new reasonings from great scientists and even becoming aware of new facts that were being discovered or deduced in the course of those 11 years of study and in those 7 years of my theory's life. To a certain extent, I was tapering toward defining a more precise timeframe, especially when I became aware of the Herto (*Homo sapiens idaltu*) fossils, which somehow confirmed, or at least indicated, that I could be right when I propose that, at the onset of human intelligence, a small decrease in brain size occurred.

Since *Idaltu* is very similar to modern humans and its brain is only slightly larger, according to my theory, the process had not yet begun when Herto's fossils were alive. Considering that their dating was 160,000 years ago, **my new timespan for the process of the emergence of human intelligence jumped from 500 thousand years ago to 160–40**

thousand years ago. I arrived at a time period of 120 thousand years. I think that is a reasonable time, and it's not incompatible with my ideas at all. My theory coexists well with the process starting even 500,000 years ago and completing 40,000 years ago. Nonetheless, I always felt that everything happened much faster, and so I continued to search to determine an even more precise timeframe.

I never imagined that I would be so precise as to determine the day it all began. If I had to bet on a date for the beginning of the human intelligence emergence process, I would put all my bets on 75,000 years ago, the day that the Toba super volcano erupted on Sumatra Island in Indonesia.

Part II

Toba

8

An unimaginable volcano

Situated on the island of Sumatra in Indonesia, Toba is currently the largest volcanic lake in the world. That is, it probably is, since its volcanic origins have yet to be definitively proven. There are, however, many facts that lead to this conclusion. Wikipedia, in a search for "Lake Toba" showed the following from 2010 to February 2015:

> "In 1949, Dutch geologist Rein van Bemmelen claimed that Lake Toba was surrounded by a layer of ignimbrite rocks, and that its origin was a large volcanic caldera. Subsequent investigators have found rhyolite ash, similar to ignimbrita, around Lake Toba, in Malaysia, and India, at a distance of 3,000 km. Some oceanographers discovered ash from Lake Toba at the bottom of the eastern Indian Ocean and the Gulf of Bengal."

But Lake Toba was already famous, at least in the scientific community, due to the Toba Catastrophe Theory proposed by the anthropologist Stanley H. Ambrose of the University of Illinois at Urbana-Champaign. He proposed that the lake was formed by a mega eruption of the Toba Volcano between 70,000–78,000 years ago (for practical reasons, I will use 75,000 years in this book), which would have been three thousand times greater than that of Mount St. Helens in the United States, and would have opened a crater 100 km long and 30 km wide. In a rating ranging from 0 to 8, the Toba eruption would have reached level 8: mega-colossal! This eruption was the most powerful of the last two million years, and perhaps even the last 450 million years (KLEIN, Richard G., BLAKE, Edgar. *The Dawn of Human Culture*, p. 223).

On the Internet, Google Maps, or Google Earth, Lake Toba is easily visibly in the island of Sumatra, Indonesia. Figure 11 is a screenshot of the lake image as seen in Google Maps.

Imagine a volcano with a crater occupying an area of 3,000 km2 (corresponding to 300,000 soccer fields), with a column of fire, smoke, ash, dust ,and gases reaching a height of 30 km! For comparisons, commercial jets travel at an altitude of 7 km, and an eruption, to influence the overall climate of the planet, needs to reach 15 km in height. Therefore, Toba would have reached more than four times the altitude of commercial jets and two times the height needed to influence the Earth's climate. The duration of the eruption is estimated at two weeks and would have expelled 670 km3 of magma. An astonishing amount of ashes covered India, Pakistan, Thailand, southern China, and the Persian Gulf region in a 3–5 cm layer and spread all over the planet. Vestiges are found today in various parts of the world,

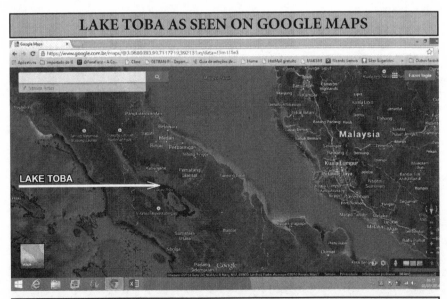

LAKE TOBA AS SEEN ON GOOGLE MAPS

Note: Screenshot (print), from Google Maps (www.google.com.br/maps). The arrow and the name LAKE TOBA were added by the author to facilitate the location of the volcanic lake on Sumatra Island, Indonesia, South Asia.

Figure 11

including in the glaciers of Greenland, practically on the other side of the globe. Due to the absence of books on the subject, I collected this data from various Internet texts, including Wikipedia, by searching for "The Toba catastrophe theory" and "Lake Toba."

In any case, what seems certain is that the theory of Stanley H. Ambrose has been well received. As much as I searched the Internet, I could find nothing to counter this theory. On the contrary, the general opinion is that it makes much sense, and the evidence found all over the planet points out that this event actually happened in a way that is at least similar to what is proposed in the Toba Catastrophe Theory.

The effects of such an explosion would have been absurdly catastrophic. The direct and immediate consequence of the eruption may have led to the annihilation of almost all living beings that were proximal to the event. Since the ashes covered India, Pakistan, Thailand, South China, and the Persian Gulf region in a 3–5 cm-thick layer, I imagine that the vast majority of living beings inhabiting the area covered by a circumference with the center in Toba and radius the size of the distance from Toba to the Persian Gulf died in the first month after the event. In Figure 12, I present a map to give the reader a clearer view of the position of the Toba Volcano on the planet.

The substances thrown into the atmosphere, especially smoke, dust, ash, and gas, blocked the sun's rays for six years, causing an ice age or anticipating an ice age that lasted for about 1,800 years of volcanic winter. This would have lowered the global temperature by 15 degrees in the first six years and by 5–12 degrees in the following years. According to most of the texts I have read about Toba, humankind has never come so close to extinction. Severe aridity seems to have drastically reduced the number of hominids for about 60,000 years on the African continent (KLEIN, Richard G., BLAKE, Edgar. *The Dawn of Human Culture*, p. 23).

Between 100,000 and 50,000 years ago, Neanderthals occupied Europe and West Asia. A number of anatomically modern hominids, much like us, were occupying Africa, and other hominids, not similar to us or the Neanderthals, lived in East Asia (DIAMOND, Jared. *The Third Chimpanzee*, p. 57). We were quite branched out. I believe that if the eruption of Toba had not occurred, and nature had somehow engendered another way of endowing us with human intelligence, we would have many animals in zoos that were

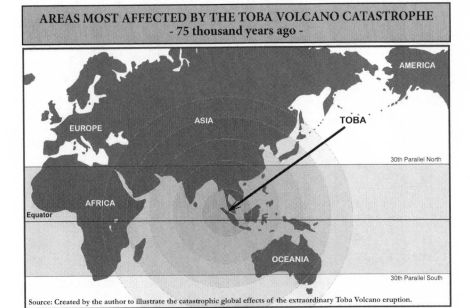

AREAS MOST AFFECTED BY THE TOBA VOLCANO CATASTROPHE
- 75 thousand years ago -

Source: Created by the author to illustrate the catastrophic global effects of the extraordinary Toba Volcano eruption.

The grey-toned circumferences denote the areas most affected in the first 30 days of the explosion, and a part of the lightest grey tone between the two parallels, 30 North and 30 South, refers to an area in which the volcanic winter, which occurred in the following millenia, caused the extinction of the greatest number of species.

Figure 12

similar to us in features but lacking in human intelligence.

In Figure 13, I display a graph to provide a closer view of hominids living on the planet 2.1 million years ago. It is a detail of the chart from Figure 4 from Chapter 1, which covered a greater timespan of 7 million years. It can also be seen as a stage for the two graphs of Figure 14: the first, an even greater detailing of the graph of Chapter 1, now covering a time from 350 thousand years, and the second, with the same time reference, a presentation of my suggestion for the Toba strand of my Human Intelligence Emergence Theory. I will explain this chart in subsequent chapters.

According to Stanley H. Ambrose himself, between 1,000 and 10,000 Homo sapiens females would have survived the super volcano catastrophe. Other scientists are far more drastic in their conclusions:

Harpending H. C. et al., 1993: 40 to 600 females

Rogers A. R., 1993: 500 to 3,000 females

Ayala F. J., 1996: 1,000 to 4,300 females

Takahata N. et al., 1995: 1,000 to 4,300 females

Just to work with a number that are somehow represents unanimous or less divergent, I did a weighted arithmetic mean and arrived at 2,574 females (rounding: 2,500 females).

When I read about Toba for the first time, I found it odd that one could make assumptions about facts that had occurred so long ago. But here and there, I translated some texts in English with the help of computer translation, and I realized that the proofs make sense, are logical, and some support the others in support of the Ambrose theory. Genetic studies dating to the 1990s, based mainly on the very low genetic variation of modern humans, for example, point to a genetic bottleneck in the human population around 70,000 years ago that reduced the population to approximately 10,000 individuals.

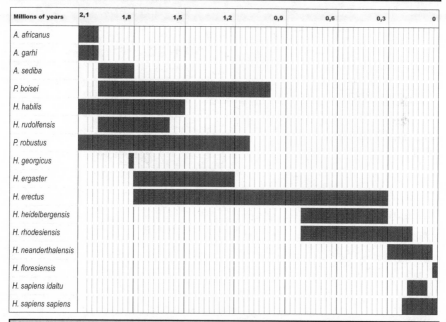

Figure 13

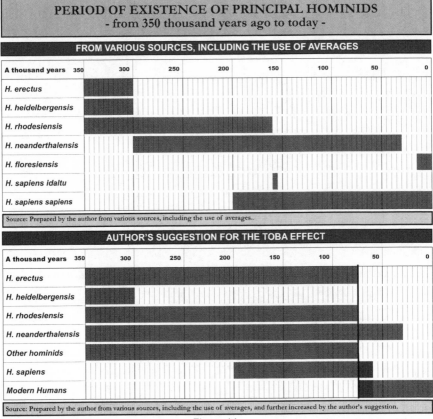

Figure 14

The genetic bottleneck effect is an evolutionary event that causes a sudden decrease in the population of a species. It can be triggered by several reasons, including a natural catastrophe. Thus, the genetic study supports the Toba Catastrophe Theory, as it predicts a bottleneck around the same time as when the volcano's possible mega eruption.

Another study that supports the Ambrose theory is that of anthropologist Alan Rogers, a professor at the University of Utah, on human parasites. He argues that the record of our past is written in our parasites. By analyzing lice genes, he confirmed that the population of *Homo sapiens* (or modern humans, as I call them) grew remarkably after the migration of a small group of humans from Africa to Asia some 150,000–50,000 years ago.

There are studies that also show that there was a bottleneck effect at the time of the great Toba eruption within the populations of chimpanzees, gorillas, leopards, rhesus macaques, orangutans, and tigers. This also reinforces the evidence that an extraordinary event occurred during this period. It should be noted, however, that the bottleneck effect was much more pronounced in hominids than in any other group of so-called higher animals, presumably because they were more vulnerable and less prepared to deal with the phenomena that occurred.

I had already found *Homo sapiens idaltu* by searching on the Internet. Then, I found Toba. That is when I thought: the effects of the Toba explosion may have somehow forced a decrease in the opening of the pelvis of a small number of humans left somewhere, probably in Africa, the continent that is considered by most authors which I have read as the most probable place for the emergence of mankind. **Here might be the accelerator of cognitive changes. It was very coincidental that a fact as extraordinary as the Toba explosion dating to about 75,000 years ago had occurred at an intermediate time between the dating of the Idaltu fossils 160,000 years ago and the first guaranteed traces of intelligence 40,000 years ago.** It is logical to presume that there was a relationship there. *Homo sapiens idaltu*, with a slightly larger brain than *Homo sapiens*, had revealed to me that I might be right when I thought that the emergence of human intelligence should have begun a little closer to 40,000 years ago, when the first proven traces of an intelligence like that of modern humans appeared in fossil records. The Toba Catastrophe Theory showed me when the process might have begun. To give substance to my idea, I needed to explain how and where.

Considering the Toba eruption, dating to 75,000 years ago, as the initiator of the process, and traces of human intelligence appearing 40,000 years ago, complex human cognition would have been completed in 35,000 years. By evolutionary standards, I recognize that apparently this is a very short period, but one must take into account the formidable evolutionary pressure that the catastrophe would have had on several animals, especially hominids. It would have been such an overwhelming force that it caused the extinction of several species, especially those living in the tropical zone of Asia, Africa, and Oceania. Moreover, any mutation or attribute that would provide a survival advantage in the new scenario would spread much more rapidly because of the small number of survivors most likely living in one place, which would accelerate the formation and fixation of new qualities. Further on, I will try to demonstrate how human intelligence took less than 35,000 years to develop.

One of the reasons we have little variation at the genetic level is because we lost it at some point, and it may be one of the consequences of the bottleneck effect. The example in the DVD *How We Became Humans* (Disc 3), shown by the geneticist Spencer Wells (29 min), is both interesting and enlightening. It shows a pot full of bullets of various colors to represent the initial population and removes some of them to symbolize the population remaining after the bottleneck. One clearly sees that many colors, representative of the genetic variations, were lost forever. Chimpanzees, gorillas, and orangutans are three–four times more genetically diverse than modern humans, indicating that the Toba Volcano catastrophe was far less harmful to them than to hominids 75,000 years ago. This shows that these hominids were very vulnerable to drastic climate change. But some escaped the effects of Toba! I think the place where they lived was of fundamental importance to this survival. **By my logic, some of those who lived in the extreme south of Africa escaped, more precisely, those in the vicinity of the sea.**

The reader who looks closely at the map in Figure 12 will see that Lake Toba is near the equator. Thus, discounting the effects of the winds, which we do not know whether they can be properly calculated for such distant dates, it is presumed that the eruption of Toba in the first days should have hit the tropical zone of the planet, especially the area near it, spreading volcanic material throughout the region and drastically lowering the temperature by 15 degrees, destroying almost all vegetation, killing most terrestrial animals, and leaving those who escaped starving. I think that in the tropical zone, from the extreme west of Africa to the extreme east of Oceania, those hominids who escaped the direct consequences of the first two weeks of eruption and the first month of famine and cold probably died in the six weeks that decimated several plants and animals in the food chain. At that time, evidence of a bottleneck phenomenon in chimpanzees, gorillas, leopards, rhesus monkeys, orangutans, tigers, and humans clearly points to this. I imagine that aquatic animals were more likely to survive, but I didn't find any information on this.

At this point in my thinking, the question that arises is: how did some hominids escape and why? The only parameter I have to analyze are the hominids themselves, as I didn't obtain more detailed data for the other animals that went extinct. Common points and similarities always help with this type of investigation. It is certain from the fossil record that there were several hominids at that time (before the eruption of Toba) living in both Africa and Asia, but also, according to fossil records, there were only *Homo sapiens* in Africa, *Homo neanderthalensis* in Europe and Asia, *Homo erectus* (a single fossil) on East Java Island, Indonesia, and *Homo floresiensis* on Flores Island, also in Indonesia.

The world map in Figure 12 shows that Europe, where the *Neanderthals* lived, is a little above 30th parallel north, which has the same distance from the Equator as the 30th parallel south, in the extreme south of Africa. If the hominids (*Homo neanderthalensis*) that lived above 30th parallel north, where the tropical zone ends and the temperate zone begins, survived that is, in a cold environment, one must think that also in the South those hominids (*Homo sapiens*) may have survived that lived below the equidistant 30th parallel north, where the tropical zone ends and the temperate zone begins, also a colder environment. Thus, the logical deduction is that the *Homo sapiens* who escaped were those most adapted to the cold, probably those who lived in the extreme south of Africa, present-day Republic of South Africa.

What about *Homo erectus*? Only one *Homo erectus* fossil that proved to have resisted Toba has been found: on the island of Java, Indonesia, dating to 50,000 years ago. Consequently, several possibilities may invalidate that proof. As it is unique, there is always more chance of error, for example, of being a *sapiens* with some deformation. Even it has been correctly classified, it may be from an ancestor of some *Erectus* who escaped in North Asia, also above 30th parallel north. Well adapted to extremely cold climates, it survived the rigorous volcanic winter and then descended to Java Island, Indonesia, where it died and where its fossils were found later. As for *Homo floresiensis*, I have already said that I ascribe to the thesis that it is a modern human who underwent the phenomenon of island dwarfism, but, even if it is an *Erectus*, it poses no problems for my theory. It could also be descended from some Erectus that was also to the north of Asia, above 30th parallel north, at the time of the eruption of Toba.

Thus, I consider the place where the hominid lived the main condition to enable him to escape the Toba catastrophe. The point common between

Neanderthals and the ancestors of modern humans, which I propose have survived, is just this: they already lived in a cold climate and were also accustomed to this kind of inclement weather. What this situation would have brought about is really interesting: in fact, it is quite fantastic!

I shall now try to explain how the events proceeded according to my theory, which regards the eruption of Toba as the initiating event in the process of the emergence of human intelligence or the emergence of humanity.

I start by sharing an opinion about the *Neanderthals*, about how they should have been at the time of the Toba catastrophe. According to general thinking, 500,000 years ago, the ancestor of the *Neanderthals*, probably *Homo heidelbergensis*, went up to Europe, while our ancestor, also *Homo heidelbergensis*, remained in Africa (KLEIN, Richard G., BLAKE, Edgar. *The Dawn of Human Culture*, p. 154). Thus, it seems logical to assume that *Homo heidelbergensis* were the ancestors of both the *Neanderthals* in Europe and *Homo sapiens* in Africa.

I should clarify that my view of how *Neanderthals* developed in Europe differs completely from the view widely accepted by science. Most of the texts I have read about *Neanderthals* regard them as having had special cognition, something between that of modern humans and other animals. Some authors admit, and even claim, that they already wore garments and ornaments and even performed funerals. Nevertheless, the stone tools they produced were practically the same as those of *Homo sapiens* before the emergence of human intelligence. I see, therefore, no reason to suppose a kind of cognition at least similar to that of modern humans.

The alleged evidence of superior intelligence found at *Neanderthal* sites does not convince me. I think that what happened to the *Neanderthals'* *Heidelbergensis* ancestors is what usually happens with animals migrating from temperate to cold climates: they increase the fatty tissue of the skin, grow and thicken their body hair to help insulation against the cold, as well as make other similar transformations that are unfortunately not preserved in fossil records or detected by today's technologies. From the shape of the bones found, the *Neanderthals* are supposed to have been much sturdier than modern humans, but the fossil record does not preserve their hair, and my opinion remains only thus. But to say that the *Neanderthals* had hair as thin as modern humans, as is now accepted by science, as the drawings of these hominids demonstrate, is also reduced to an opinion to the best of my knowledge; as there is no evidence that they had thick, long hair, there is also no evidence that they were bare like us modern humans. But I take as a basis a proven logic that animals grow and thicken their hair when they change from temperate to cooler climates. A classic example of this phenomenon is the woolly mammoth, which, let's say facetiously, is a fatter elephant with a fur coat.

Because of this, I think *Neanderthals* at the time of the Toba eruption would have had a physical form resembling the "Abominable Snowman," a legendary and mythical figure who supposedly lives or lived in the Himalayan mountain range, where Everest, the highest mountain in the world, reaches 8,848 meters of altitude.

Even when they lived outside of caves, the *Neanderthals* left no convincing evidence that they could build solid houses, even though they had to face extremely cold weather conditions (KLEIN, Richard G., BLAKE, Edgar. *The Dawn of Human Culture*, p. 159). One might even say that I am exaggerating, and Europe was not and is not so cold. I agree, but I think that the *Neanderthal* group that managed to survive Toba was the one that lived in the northernmost part of Europe, and therefore was much more adapted to cold and long winters. Perhaps he had even developed a rudimentary system, say, of hibernation, which would make it easier to survive the volcanic winter caused by Toba.

So I think this explains the survival of the *Neanderthals*: they were better prepared to face a volcanic winter and even an Ice Age. It had nothing to do with special cognition. This is not my prejudice against the *Neanderthals*. After all, I think that, up to 75,000 years ago, our ancestors, the *Homo sapiens* living in the extreme south of Africa, like the *Neanderthals* and all the other so-called superior animals, also had a similar cognition that was nothing special.

These *Sapiens*, our ancestors—why did they escape? In the early days, certainly because they lived in a fairly cold climate, they were also better prepared to face the catastrophe. But they did not have the apparatus of the *Neanderthals* to withstand the rigors of future winters and probably no surviving large animals were left in southern Europe for them to eat. After all, Africa ends in the South at a point that is approximately at the same distance from the Equator as is the point where Europe begins in the North.

The *Neanderthals* still had northern Europe to which they could escape. The *Sapiens* were at a crossroads. Southern Africa ends in South Africa. After the Toba eruption, food was missing, and these hominids were stuck without the slightest chance of escape because the sea was to the east, west, and south and the icy hell left by the Toba explosion was to the north. Climbing to the temperate zone meant a certain death. To go further south was to enter the sea. We had never been so close to extinction! My idea is that nature's attempt, or attempts, to solve the question instigated the emergence of human intelligence there in South Africa, from where modern humans, now, yes, that we can call humanity, spread all over the planet.

At this point, my thesis is clearly against the Multiregional Theory, which argues that modern humans evolved in parallel in different parts of the world. My theory, regardless of whether or not Toba was the cause of the unleashing of the process of the emergence of human intelligence, is consistent with the Single Origin Theory, which argues that modern humans descended from a group of hominids that have left Africa 60,000–55,000 years ago and spread throughout the world. I consider the facts that resulted in the emergence of human intelligence to be so extraordinary that I imagine it has occurred only once in the whole universe. Therefore, until the moment when I finish this book, I do not believe that there are, at least with the rules to which I have access, extraterrestrial beings with intelligence similar or superior to ours.

Now, I can better explain the graph of Figure 14, where I suggest that *Homo rhodesiensis, Homo erectus, Homo neanderthalensis, Homo sapiens,* and others lived on the planet at the time of the Toba explosion. *Homo floresiensis* does not appear on this graph because, as I mentioned earlier, I consider it to be within the *Homo sapien* or the modern human category; only, it suffered from island dwarfism. The *Neanderthals* escaped the tragedy, but became extinct 28,000 years ago. If any other hominid survived, it somehow became extinct later, leaving only modern humans on the planet.

Finally, the dark part, both the ranges of *Homo sapiens'* existence and modern humans's existence refer to the timespan of transition from one species to another, at which point mankind emerged according to the Toba strand of the Human Intelligence Emergence Theory.

9

Swimming to the future

The question left hanging at the end of the previous chapter was this: what exactly happened in the extreme south of Africa that brought forth human intelligence? I will explain. Some ecological emergency was important enough to diminish the opening of the pelvis, slightly reduce the size of the brain, and cause our ancestors to be born more and more prematurely.

What would this emergency be? Here, I would like to pay tribute to Elaine Morgan, Mac Westenhofer, and Alister Hardy, who defend the idea that, during a fossil hiatus—more than five million years ago—the ancestors of hominids went through a semiaquatic period of existence before returning to a style of predominantly terrestrial life. **I imagine that the strongest candidate to cause pelvic narrowing was the extreme need for swimming, in order to obtain food from the sea, pressured by the terrible icy hell into which almost the entire African Continent was transformed.** I understand that, with this, I am transposing the idea of the aquatic ape to five million years later, and I hope that the authors and proponents of the theory are not bothered by it. I also understand that the timeframe I propose seems too small for so many changes that would have occurred along with the emergence of human intelligence. But all of this needs to be viewed as a whole and as evolutionary emergencies driven by the ecological urgency affecting a small group of surviving hominids (5,000 males and females), within which all successful changes spread with extraordinary rapidity. Although I consider swimming for hunting as the main driving factor in the narrowing of the pelvis, I remember that several facts demonstrate that the endurance race hunt, still practiced today by the Sam peoples in the Kalahari Desert (MORGAN, Elaine. *The Aquatic Ape Hypothesis*, p. 49), also concurred in the same sense.

I stand by the idea that, when certain things occur, evolution accelerates to incredible rates. I do not disagree with the concepts of the molecular clock, a tool proposed in 1962 by chemist Linus Pauling and biologist Emile Zuckerkandl, which has proven to have contributed to the study of the evolution of living things. Nor do I have the technical conditions to do so. But the method, although quite reliable in its concepts, depends on the use of variables such as mutation rates, for example, most often calculated using external data and calibration techniques, which compromise the accuracy of the results. The geneticists themselves recognize this imprecision, especially when they involve long time periods. In the case of catastrophes, I envisage a randomness of change, including temporal change. Many mutations get lost, leaving no trace, and others have their adaptive success governed by sudden alterations in the environment. I estimate that ecological disasters that put a species in danger of extinction create genetic bottlenecks that disorganize their rates of mutations in future studies because they cause extraordinarily rapid changes, giving the impression that more time has passed than actually occurred.

Let us imagine a simple hypothetical example of new features that quickly develop in reclusive populations. A species of 100,000 individuals lives on an island. They only eat a fruit that sprouts in the hollow of a tree, the diameter of which is 5 cm. Climate change causes all fruits to sprout in 2-cm holes, when only 1% of the population has snouts capable of reaching fruit in 2-cm holes. In only one season, 99% of the population will disappear and, with it, all the mutations and characteristics not yet passed on to individuals whose snouts can reach the bottom of 2-cm hollows, leaving only 1,000 individuals of the initial 100,000. Thus, a drastic change in the environment can also lead to a drastic change in the characteristics of a population, which, as in the example, may tend to grow and return to its original size due to the large supply of food. In a short time in evolutionary terms, therefore, the profile of a population can undergo a radical change. Situations like these are rare but they do happen. When they do, they bring about immediate results in genetic terms that are not always easy to explain in the future.

I suggest that, after the initial days of the Toba eruption, due to which all the hominids living in the tropical climate of Africa, Asia, Oceania, and a small part of Europe became extinct, only a small number of Neanderthals in Northern Europe and a few hundred Homo sapiens living in the extreme south of Africa escaped extinction, and that we, modern humans, descended from those hominids who survived.

But I am not the only one who thinks that all humans are descended from a people who lived in South Africa. Archaeologist Curtis W. Marean, a professor at the School of Human Evolution and Social Change at Arizona State University and an associate of the Institute of Human Origins, has developed a theory that modern humans developed precisely in southern Africa, present-day Republic of South Africa. He came to this conclusion on a path totally divergent from mine, based on an event of which I was unaware: a long glacial phase, known as Marine Isotopic Stage 6, began 195,000 years ago and lasted until 123,000 years ago. Even admitting that there is no detailed record of environmental conditions in Africa during Marine Isotopic Stage 6, with support for the most recent and best-known glacial phases, climatologists suppose that these climatic conditions were almost certainly cold and arid, the deserts were probably very larger than the present, and much of the land mass would have been uninhabitable. To summarize, a cold and dry ice age had desertified much of Africa, making almost the entire continent uninhabitable. Marean proposes that in the extreme south of Africa, in a cove called Pinnacle Point, where he made excavations with surprising results, lived a group of our ancestors, who, driven by the new climatic conditions, developed complex practices to obtain their food from the sea, including crustaceans, fish, and even mammals such as seals and whales. This new and diverse way of eating, he concludes, provided the conditions for the emergence of humanity.

It even seems that we concur, because his ideas match my theory, determining a greater and therefore more appropriate period for the occurrence of the evolutionary changes that I presuppose (190,000 years ago − 123,000 years ago = 73,000 years) than with time calculated when I consider that the Toba eruption triggered the process (75,000 years ago −

40,000 years ago = 35,000 years).

But history, here with a broader definition than the traditional one assuming writing as a parameter, is that which happened and not that which seems to have happened because it satisfies conditions considered true and interesting to those who analyze it. To suit my theory, I recognize that in principle, the process beginning 195,000 years ago with a glacial period of 70,000 years is much easier to accept than a process beginning just 75,000 years ago with the mega-catastrophic eruption of Toba. However, several other facts and analyses have caused me to opt for the Sumatra volcano explanation, as I will now justify.

If the process had begun 195,000 years ago, one question remains: why do we find no trace of this gradual increase in cognition before 40,000 years ago? We cannot say that fossils have not been preserved because we are not looking for fossils. We are looking for evidence of a gradual cognitive change, and such evidence is always present in stones, which are very resistant to the weather. Without such evidence, I believe that Toba is the most logical technical solution. But I recognize that it is also the most spectacular, beautiful, and creative solution and that my vanity, which I acknowledge, may have led me to adopt it in the beginning. I spent much time thinking about completing the book without including Toba, in fear of being ridiculed, but the more technically I thought about it, the more I became convinced of its plausibility. The more I read about the extraordinary Sumatra volcano, the more it impressed me with the marrying of its ecological consequences with my theory.

Beginning 2.5 million years ago, the pelvis of our ancestors underwent three major changes. The first was when they descended from the trees and adopted an upright walk. From the myriad images I have seen of skeletons of monkeys, hominids, and modern humans, it was the position that most changed, certainly to accommodate bipedal walking and supporting the trunk organs. The second change, I believe, was progressive and continued for over a million years, a period when the brain grew and the pelvis had to keep up with that growth to ensure the delivery of that stubborn animal.

The third major modification is my proposition and is already part of the theory. So subtle, it cannot even be called great, for to this day either it was not noted or it did not matter: a slight diminution in the pelvic opening. I suggest that this third change occurred with the group of hominids which, I think, escaped the ecological disaster, settled in the extreme South of the African continent, and gave birth to all humankind. I saw in this small diminution of the opening of the pelvis the determining factor that caused the events that initiated the appearance of the human intelligence. The reasonings I have developed from there, many of which I will further explain in this book, have persuaded me to choose Toba as my preferred impetus for the Human Intelligence Emergence Theory.

I believe that *Homo sapiens* were already living in the extreme south of Africa when the Sumatra volcano erupted and that they were already feeding on crustaceans and even some fish. But I do not believe that they were already provided with a beginning of human intelligence merely because fishing for crustaceans is a complex activity that requires a privileged cognition, an idea defended by Marean. As I have noted above, many complex activities are performed by animals absolutely devoid of modern human-like intelligence.

I think that if, before Toba, some hominids already consumed crustaceans and fish, these were an extra food source. The nutritional base on the continent continued to be animals and terrestrial vegetables. I also think that those *Homo sapiens* were no smarter than a chimpanzee, an architect-bird from Papua New Guinea and Australia, a New Caledonian crow, a *Homo neanderthalensis*, or a nail monkey here in Piauí. Bird-architects, for example, construct arbors so complex that they seem to be made by modern humans, with the sole aim of seducing the largest number of females (DIAMOND, Jared. *The Third Chimpanzee*, pp. 192 and 193). New Caledonian crows make tools, finding exactly the right branches, breaking them properly, twisting their tips, and making sharp hooks to extract worms, insects, and other invertebrates from deep crevices, according to studies, field and laboratory observations by Russell D. Gray, Gavin Hunt and Jennifer C. Holzhaider of the University of Auckland (New Zealand) and Anne Clark of the State University of New York in Binghamton. The fact that hominids produced no more than a single stone tool for 2.5 million years, with very little diversification, seems to indicate that they did not have an intelligence different from most other animals.

This is where I perceive some incompatibility with the process beginning 195,000 years ago or more. It would be as if a new cognition was emerging, but it did not leave any external and lasting trace such as through drawings, paintings, and sculptures. The tools, which logically would have been the first to be affected, practically did not change until 40,000 years ago. The supposed proofs of higher cognition before 40,000 years are more conjecture than fact and always with dating limited to 70,000 years ago. Since there is never a consensus when it comes to dating, 70,000 years ago may well be read as 90,000 years ago or as 50,000 years ago, depending on many factors, including choice of the dating method or trusting the accuracy of the layer where the material was found.

Recently, in early 2013, a controversy began about the dating of the Neanderthals. A new dating method called ultrafiltration suggests that these hominids became extinct at least 50,000 years ago, which would rule out any possibility of an encounter with modern humans, who would have reached Europe 35,000 years ago. Polemics about dating exist all over the planet because a kind of dispute has been established for older fossils and lost links that invariably lead to the confrontation of irreducible positions and the presentation of new methods of evaluation, often with surprising results.

Dates, therefore, are grounds for controversy even within the same research team. In addition, evidence related to human cognition before 50,000 years almost always refers to ocher supposedly used as painting, manipulated objects also supposedly used as adornment, and traces of the use of fire.

With regard to the conquest of fire, I have an absolutely different view of that which is now accepted by science. For me, fire control is a very complicated activity to have been performed by a living being devoid of human intelligence. I believe that only modern humans truly managed to control fire, and that the traces found before 50,000 years ago originate from accidental fires, because accidental fire is something that has always been present on our planet. However, some authors, such as Richard G. Klein and Black Edgard, argue that there is evidence of the use of fire by humans 250,000 years ago (KLEIN, Richard G., BLAKE, Edgar. *The Dawn of Human Culture*, p. 131). Richard Wrangham says that the use of fire is much older, dating back to about 1.5 million years ago and supports the idea that only the cooking of food would have converted *Homo habilis* into *Homo erectus* (WRANGHAM, Richard. *Catching Fire - How Cooking Made Us Human*, p. 84). I've read what I could find on this evidence, but I remain convinced that a being with a lower intelligence than that of modern humans could not keep a fire burning properly in order to use it to cook food. It is such a complex activity that even modern humans cannot carry it out easily, either letting the flame go out or causing some disaster for themselves, other humans, or the place where they are cooking.

Perhaps the environment most conducive to the development of fire control has been the beaches close to where I think these hominids lived in the extreme south of Africa, as this would greatly reduce the risk of uncontrollable fires, thus facilitating learning. Based on this understanding, I imagine that humans started to control fire after the Toba explosion, most likely around 65,000 years ago.

I believe that, a few days after the eruption of the Sumatra volcano, that handful of *Homo sapiens* living in the extreme south of Africa began to face perhaps the greatest food crisis of the *Homo* genus. And here arises the great difference in the situation of the *Neanderthals* and of the hominids during the Toba catastrophe: in Europe, they could flee northward; in the extreme south of Africa, as I said before, they were trapped and had only one way to survive: by taking food from the sea. The food from the sea, now with the volcanic winter, would no longer be extra food: it would their only food source for no one knew how long. Swimming for hunting and/or endurance racing would have become an evolutionary urgency. A small advantage in the art of swimming and running, the decrease in pelvic aperture, for example, as I argue, would have made the difference between life and death. Premature birth, with a slightly smaller human brain, became a decisive feature for survival. This fits the great eruption of Toba into my theory of prematurity and in my Human Intelligence Emergence Theory in a way that I think is quite convincing.

According to this understanding, I believe that the extraordinary ecological problem caused by the mega eruption of Toba brought about an extremely urgent reduction of the pelvis opening, which caused prematurity to allow for childbirth, which caused those *Homo sapiens* to be born with more free memories, that were used by algorithms such as language, comparison, imitation, counting, differentiation of sounds, recognition of group members, recognition of other animals, multiplication, division, addition, proportionality, and many others. It also fostered an environment conducive to the emergence of other absolutely new brain algorithms in an inflationary process, driven both by the now natural advantage of being smarter and by the natural advantage of having a narrower pelvis for better swimming and taking food from the sea. The emergence of human intelligence would have been a chance occurrence involving so many variables that it will hardly have occurred or will occur anywhere in the universe. This probably makes us unique and attempts to contact extraterrestrials absolutely unsuccessful. It would be better if the time spent on these projects were better applied to ensure the survival of the extraordinary being that we are, despite our many defects, including moral and ethical ones. Or, principally moral and ethical ones, as the reader may prefer.

I believe that in such a situation, driven and fueled by an external factor such as prematurity, something like what Richard Dawkins called "software and hardware coevolution" may actually have occurred, with software and hardware innovations pushing each other in a growing spiral. It is logical to expect that this new situation of free memories and new algorithms would have caused a new increase in the childhood period of humans, especially to permit adaptation of the new cognitive resources to the way of life of those *Homo sapiens*. This fits in with reality, for we are the primates with the longest childhood.

I think that starting with Toba, we began to become more intelligent, a process which I believe continues to this day, perhaps now also for other reasons that need to be duly studied and clarified in a freer scientific environment less dependent on political decisions, which place a smokescreen on truths that humanity insists on not seeing.

At this point it is good to clarify one question: which clusters of *Homo sapiens* escaped in the extreme south of Africa and gave rise to modern humans? I consider as potential candidates the hominids who lived in the archaeological sites of Pinnacle Point, Blombos, and Foz do Rio Klasies, dating respectively from 170,000 to 40,000 years ago, from 140,000 to 71,000 years ago, and from 120,000 to 60,000 years ago. The reader should note that dates older than 75,000 years do not preclude the idea that Toba's mega-explosion was the starting point of the emergence of human intelligence. As I have already mentioned, hominids already living in the extreme south of Africa were the ones who escaped the Toba catastrophe. Thus, dating older than 75,000 years only confirms that there were hominids in the region, and dating after 75,000 years shows that these hominids escaped the Asian volcano catastrophe. The idea that Toba was what caused the process is at risk if it is found beyond a shadow of a doubt that there were hominids with the behavior of modern humans more than 75,000 years ago. To the best of my knowledge, there is no undeniable evidence of this, neither in South Africa nor anywhere in the world.

Before closing this chapter, I wish to point out that it was in Blombos that the team of Christopher Henshilwood, a professor and researcher at the University of the Witwatersrand, found in a layer that indicates a dating around 70,000 years ago the Stone of Blombos—the most ancient evidence of geometric thinking, the image of which I have chosen for the cover of this book.

10

The exiguity of the time

At this point in the reasoning, I think I have explained my idea of the emergence of human intelligence, demonstrating that this process may have occurred sometime between at least 500,000 and 40,000 years ago and most likely between 195,000 and 40,000 years ago. That is, I explained the logic of events according to the principles of my Human Intelligence Emergence Theory, which is based on the premise that some extraordinary event caused free memories to appear in the human brain, which were used both by existing algorithms and by others that arose, taking advantage of the new evolutionary environment that this situation had fostered. This book was to end at that point. I did not intend to go any further, until I learned of the mega-eruption of a super volcano on Sumatra Island, Indonesia, 75,000 years ago and that this fact could have occasioned a genetic bottleneck effect in various animals including humans as advocated by the Ambrose's Toba Catastrophe Theory.

These facts changed the structure of this book, because the Sumatra volcano explosion and the Ambrose theory fit perfectly into my theory. But the incorporation of Toba brought me problems because in the light of evolution, the time left for the emergence of humanity turned out to be very less. If the great eruption of Toba occurred 75,000 years ago and the first signs of intelligence arose 40,000 years ago, we had 35,000 years for a complete development that transformed a chimpanzee-like cognition into a cognition like that of a modern human, when life has existed on Earth for 3.5

billion years; the first animals for 1.2 billion years; mollusks and trilobites for 542 million years; the first fish for 488 million years; modern fish for 416 million years; the first insects for 318 million years; the first dinosaurs for 251 million years; the first placental mammals for 150 million years; the first large mammals for 65.5 million years; the first dogs, elephants, whales, and bats for 56 million years; the first rhinoceroses, cats, camels, horses, great apes, and bears for 23 million years; and the first *Australopithecines* for 5 million years. None of them developed anything like human cognition. Not even the dinosaurs, who dominated the planet from 251 to 65 million years ago, when they became extinct, left a single simple mark on some rock.

I confess that it is difficult to accept such a short time for such a complex event that certainly involves cognitive changes never before used. However, recent studies have shown that the occurrence of a mutation in a gene, which is a code for a transcription factor, may cause its expression to alter the expression of several genes, leading to a tiny change in a promoter of the DNA that controls the transcription factor to cause a cascade of differences to the body, enough to create a new species without completely altering the genes themselves. (RIDLEY, Matt. *Nature via Nurture*, p. 50). DNA promoters express themselves in the fourth dimension: time defines the changes. Just as a chimpanzee has a head different from a human's, not because it has a different program for the head, but because it develops the jaws for longer and the skull for less time, modern humans may have acquired brains prone to human intelligence not because they have a different Homo sapien program for the brain but because their free memory spaces have been developed for longer and other parts of the brain for less time.

But can humans, depending on environmental conditions, also quickly develop other remarkable features? It seems so, as is the case with lactose tolerance, systematically documented by William Durham in his book Coevolution (DAWKINS, Richard. *The Ancestor's Tale*, pp. 51 and 52). Human babies are born with the lactase enzyme activated in order to process lactose, the sugar found in milk. From a certain age, between 2 and 5 years in some modern humans, a deactivation of this enzyme takes place, causing the affected humans to no longer be able to digest the carbohydrate. These humans, when they drink fresh milk, feel ill and suffer intestinal cramps and diarrhea.

However, other humans are able, in adulthood, to normally consume milk without the slightest health problem. The explanation is that this happened after some groups of humans began to domesticate recently, in evolutionary terms, perhaps 5,000 or 10,000 years ago, animals like sheep, goats, and cattle, including using the milk of these animals to feed their babies. It is easy to imagine that an evolutionary pressure had been established in these humans to keep lactase functioning even after childhood, especially if there had been a long period of difficulty in obtaining food, as is always the case with several species, sometimes because of climatic phenomena, which generally affect plant food.

But why are not all humans be able to digest lactose? Because not all the human groups around the world started to raise sheep, goats, and cattle, and because of this, they did not undergo evolutionary pressure to digest milk after 3–5 years of age. Thus, some human groups—Chinese; Japanese; Inuit, most Native Americans; Javanese; Fijians; Australian Aborigines; Iranians; Lebanese; Turks; Tamils; Sinhalese; Tunisians; and many African tribes—including the Sans, Tswuanas, Zulus, Shosas, and Swasi of Southern Africa; the Dinkas and Nueres of North Africa; and the Yorubas and Igbos of Western Africa—do not do well with milk after that age. In general, these peoples do not have a history of pastoralism. The rest of mankind, however, consumes milk during all periods of life without any problems. These facts point to a very recent genetic change, most likely brought about by our own domestication. We humans have also been domesticated. In fact, we were the first beings to be domesticated by ourselves, as we continue to do until today. What we call educating is, in fact, domesticating.

However, still more fossil evidence emerged to further aggravate the problem of time. According to recent datings, Australian aborigines arrived in Oceania 50,000 years ago (DIAMOND, Jared. *The Third Chimpanzee*, pp. 299 and 372). Thus, to admit that mankind only arose 40,000 years ago will almost say that Australian aborigines, who arrived 10,000 years earlier in Oceania and never had contact with other humans for genetic exchanges, since they were only contacted by Europeans in the fifteenth century, were not yet modern humans, which seems to be impossible. Consequently, my time is shortened by 10,000 years, and the most likely interval of completion of the modifications necessary to move from an animal with a chimpanzee-like cognition to an animal with a cognition like that of a modern human would be 25,000 years, beginning 75,000 years ago and completing 50,000 years ago, when a group of hominids, like modern humans, ventured through Asia and ended up arriving in Oceania.

In view of this, 50,000 years ago could be the date for the complete emergence of humanity. However, modern human migration data point to an even shorter time: 60,000 years ago (more or less, in accordance with new studies and datings that will surely come), as I will elaborate further in chapter 26, **African genetic diversity**.

Even with the shorter development time, I continued and continue to consider Toba as the detonating event of the process. And I will try to defend my point of view in the coming pages, showing that many events point in this direction, especially when an extreme situation causes very rapid changes in a population suddenly reduced by a bottleneck effect, made persistent by the enduring consequences of the phenomenon that caused it: the extraordinary explosion of the Toba super volcano.

Evolution can occur very fast or very slowly, depending on several factors, many of which I do not know, and others that even today science does not know. But let me mention two very clear examples that will demonstrate how the speed of evolution is variable.

Crocodiles have been on the planet for 250 million years, always living in a similar way, partly on earth and partly in water, making them quite adaptable to environmental variations. Maybe that is the biggest reason for their long survival as a species. They lived with the dinosaurs and escaped the asteroid catastrophe that nearly wiped out life on Earth 65 million years ago, probably because they could also live in water, which seem to have been less affected by the disaster. They certainly survived calamities that affected the aquatic environment more sharply because they could live on the land as well. Except Europe and the polar regions, they are found throughout the rest of the world, and as adults they have only one natural predator: the Asian tiger. It really is a successful evolutionary project. Perhaps, therefore, they have changed little over millions of years.

Silver foxes, however, under strong evolutionary pressure, can change noticeably in a very short time. D. K. Belyaev and his colleagues captured a few specimens of these foxes and began to cross them systematically in order to obtain tame animals. By crossing the most docile foxes of each generation, in just 20 years they had obtained foxes that behaved like border collie dogs, who sought the company of humans and wagged their tails. In addition, genetically tame foxes grew black and white fur - with white spots on the face and muzzle, acquired a new hormonal balance, and began to breed all year round rather than in a specific season. They were quite similar to dogs. And note that dogs do not descend from foxes. All dogs in the world descend from wolves, more precisely from the European wolf Canis lupus (DAWKINS, Richard. *The Ancestor's Tale*, pp. 49 and 50).

If silver foxes could change so much in 20 years, to the point of altering even in the way they procreated, why then could human intelligence not have emerged in 25,000 years? Especially if it is considered, as I say, that the biggest change occurred in the increase of memory spaces and in algorithms that allowed the system to work with much more data. Moreover, evolutionary biology demonstrates that it is precisely in small populations that new characteristics are more likely to be promoted and preserved (FOLEY, Robert. *Humans before Humanity*, p. 146). So, I do not see so many problems with time. I see problems with facts of which I am unaware and which may render my theory unfeasible. But there is only one way to at least try to solve this question: by spreading my ideas, as I am doing now in this book. However, it is good to clarify that I am disclosing them and not fighting for them. I am exposing them to see if they have any meaning.

11

Human females
pay the price

If anyone paid the price for the emergence of human intelligence, according to my theory, it was the human female. I will explain why I have this opinion. In addition to the emergence of human intelligence, the narrowing of the pelvis had other consequences for humans, among them the complications of childbirth in our women is most evident. As I have already suggested, to solve the problem of passing the baby's head through the new narrow pelvis, there was a small decrease in brain size coupled with a noticeable decrease in gestation time. Since a decrease in brain size would probably lead to great losses, a decrease in gestation time was established as a solution to the problem, causing a premature birth, which caused babies to be born with slightly smaller brains, enough to pass through the pelvic canal. But some evolutionary changes, especially those that are very urgent, turn out to be traumatic, and it may have been that some human females had a narrow pelvis but not enough prematurity in childbirth for the baby's head to pass through the pelvic opening, thus dying in labor, after much suffering, often together with the child. This was until recently, when caesarean delivery technology was created. Recently, it has been found cesarean birth after death of the female has existed since the time of the ancient Egyptian civilization. However, it became popular in the last century with the advent of modern anesthesia. Before, it was only used when there was no other possible solution.

If I am right in this reasoning, women were the greatest victims of the emergence of human intelligence because, in order to avoid the extinction of our species, evolution had to narrow the pelvis, which would have caused both physical and psychological suffering that was exaggeratedly intense for human neurological patterns. Perhaps science has to explain exactly what it calls "normal" childbirth, if there even is a "normal" childbirth because, as I propose, we are still in full adaptive change, aiming to reconcile the opening of the pelvis with the size of the heads of babies, through the artifice of prematurity.

This situation has given rise to types of women who can give birth without ramifications; women who can give birth, but suffer some ramifications; women who have children with great difficulty and suffering, with many consequences; and women who cannot give birth to their children and die in childbirth, often together with the children. With the popularization of cesarean delivery, there was one more type of women: those who do not accept the certainty that they cannot give birth and instead opt for this type of delivery.

In relation to pain and suffering in birthing, as much as I study and observe, I find no animal birthing that comes close to the birthing in our females. We have perhaps the only females in the world who need help to deliver children. As far as my technical knowledge goes, few women give birth with the tranquility of other animals. Two pieces of information illustrate this well: before the onset of modern obstetrics, human females often died in childbirth, while females of gorillas and chimpanzees rarely died; a study with rhesus macaques recorded only one death in 401 observed births (DIAMOND, Jared. *The Third Chimpanzee*, p. 151). Even the women who manage to give birth to their children quite easily have complications forever. Is it right to demand so much sacrifice from our females, directing them to a risky, painful, suffering birth that can cause complications and even kill them? Karen R. Rosenberg, paleontologist, and Wenda R. Trevathan, anthropologist and biologist, in the article *"The Evolution of Human Birth,"* published in *Scientific American*, wrote the following explaining the difficulties of human birth:

"To understand the process of birth from the mother's point of view, imagine that you are about to give birth. The baby is most likely upside down, facing toward his side when his head enters the birth canal. In the middle of the canal, however, he must turn and turn to his back; the back of his head will then be pressed against his pelvic bones. At this point, therefore, the baby's shoulders are oriented to the sides of his body. When he leaves your body, he will still be turned back. This rotation helps to rotate the shoulders, so that they can also pass between your pubic bones and the coccyx. To evaluate how the maternal and fetal dimensions match, take into account that the average pelvic opening of women is 13 centimeters in its largest diameter and 10 centimeters in the smallest. The average head of a baby, in turn, is 10 centimeters from the front to the back and the shoulders are 12 centimeters wide. It is this journey through an irregularly shaped corridor that makes human birth difficult and risky for the immense majority of mothers and babies."

I do not know the percentage of women who have normal deliveries with no sequelae, those with normal deliveries with some sequelae, those with normal deliveries with many sequelae, or those who either died or had caesarean sections due to the incompatibility of the opening of the pelvis with the size of the baby's brain. Maybe there is research on this topic, but, unfortunately, I have not acquired this information. Who knows, one day I may have the opportunity to do a detailed study on the birth of modern humans, and I can obtain this information. In any case, I think these figures should be better publicized so that women can decide more safely regarding how they will deliver their children.

Cesarean delivery today is only advisable if it is found that it is impossible to give birth "normally." If my reasoning is correct, it is difficult to justify the orientation given to women today that they should try until the last minute the so-called "normal" delivery. If there is a risk of not being able to deliver the child and even die in the attempt, why direct them to approach such a dangerous risk? If there is a limit for the brain to go through the birth canal, it is logical to imagine that, close to this limit, even with a "normal" delivery, the baby's head can leave serious sequelae in the mother.

Even cesarean delivery has future implications. It is ethically correct according to the culture. One cannot let a person die suffering if there is a solution that will allow that person to live. However, by the laws of nature, the procedure is totally wrong because that female will convey to her descendants the genetic characteristic of not being able to give birth to her babies and increasing the number of women unable to give birth to their children "normally." Thus, the sacrifices of women who take the risk and who suffer and end up with sequelae have a minimal influence on the final outcome; reason points toward our having more and more human females unable to carry out their deliveries without surgical intervention, especially as more and more women choose cesarean delivery.

The reader should note that I am not advising anyone regarding how to have their children. I am not a physician, nor do I have the authority to give this kind of advice. I only set out reasoning for the purpose of doing justice to human females and to give them as much information as possible so that they have the opportunity to choose how they want to give birth to our children.

Here I note an issue that I consider pertinent to the present argument. It is brought up by paleoanthropologist Katerina Harvati of the University of Tübingen, Germany, in *"Questions for the next million years"* (Davide Castelvecchi in the special issue of *Scientific American Brazil*, Oct. 2012): *"The narrowness of the human birth canal is a major obstacle to head size. Will the continued use of cesarean sections for hundreds of thousands of years lead us to develop bigger brains?"*

I will present two answers to this question: one is based on the view of science today that the human brain needed to grow to provide humans with intelligence, but the narrow pelvis was an obstacle to this growth, and for this reason prematurity arose to solve the problem of the baby's head passing through the birth canal; the other is based on my theory, which says that the human brain grew for other reasons, and, with it, enlarged proportionally to the pelvis, until an external event caused a sudden narrowing of the pelvis, and therefore prematurity appeared to solve the problem of the baby's head passing through the birth canal. The position of science is quite clear. Before her question is the following statement: "The narrowness of the human birth canal is a major obstacle to head size." According to my theory, this statement should not be made because I consider that the head size was the obstacle to the narrowing of the pelvis.

I will start with the view of science, clearly demonstrated by paleoanthropologist Harvati, as the premise of the question. If this statement is correct, of course the use of the cesarean section, as it is today, would lead to the development of a larger brain, simply because this obstacle would be avoided. That is, for a time, no woman could give birth "normally," simply because babies' heads would grow so much they would not pass through the birth canal of any woman in the world, even imagining that this growth might have a limit. Thus, the logical conclusion is that if the general view of science on this subject is correct, probably with the onset of the use of cesarean technology, brain size may be increasing, which had ceased about 600,000 years ago, and humans must be developing a larger brains, and by today's accepted principles, becoming smarter. Unless, of course, there is no longer any pressure to increase intelligence, as is accepted that happened in the past. In this way, there would be no influence on brain size.

If I am correct, however, it changes everything. Because I say that, up to 75,000 years ago, the brain and pelvic aperture were compatible and were adequate in size for an easy delivery. What occurred was a sudden decrease in the opening of the pelvis, which caused prematurity, in order that the brain could pass through the new narrow pelvis. Why do I think the brain was not growing when human intelligence came along?

The fossils show this. According to the fossils, the human brain had not been growing for hundreds of years: *Homo rhodesiensis*, which probably lived from 600,000 to 160,000 years ago, had a brain volume of 1,300 cm3; *Homo heidelbergensis*, which lived from 800,000 to 300,000 years ago, had a cerebral volume of 1,350 cm3; *Homo neanderthalensis*, which lived from 300,000 to 29,000 years ago, had a cerebral volume of 1,450 cm3; and *Homo sapiens idaltu*, which lived from 160,000 to 100,000 years ago, had a cerebral volume of 1,450 cm3. As the fossil finds demonstrate, at least 600,000 years ago, some species of hominids with brains of a size compatible with that of modern humans, apparently with no problem with childbirth, were already roaming the planet, and some even had a slightly larger brain than modern humans, such as *Homo neanderthalensis* and *Homo sapiens idaltu*. So, I see no sign of any remarkable brain growth after 600,000 years ago. Thus, from the fossil record, it is clear that hominids' brains grew extraordinarily up to approximately 600,000 years ago and then stabilized. If there exists, as it should exist, one or more causes for the exaggerated growth of the brain, they must be sought before 600,000 years ago. Archaeological records show that the brain stopped growing for 550,000 years, and only then did the first signs of an intelligence like that of modern humans begin to appear.

According to my thesis, it is unlikely that the external factors that required the reduction of pelvic aperture would still be present in a humanity that has conquered the world and lives in various environments. Gestation time, therefore, seems to have stabilized soon after modern humans left the extreme south of Africa to conquer the world. Or it was already unchanged a little before that. Thus, cesarean delivery solves the problem that prematurity was still solving in South Africa, when the process was interrupted, as demonstrated by the fact that, according to the World Health Organization, around 15% of deliveries of human females require cesarean delivery.

If my theory is correct, science, in setting a universal datum for cesarean delivery, as it seems is being done in some way, will only increase the percentage of women who will need cesarean technology in order to have their children. There will, therefore, be no increase in brain size, although, according to fossils, the human brain has not increased for more than 600,000 years. There seems to be no evolutionary pressure occurring in that direction.

At least in principle, because of the change in the environment of natural selection caused by so many years of free memory increase and new algorithms in the brain, in case intelligence has become a decisive feature for survival, an opportunity may have arisen for a situation to exist that allows for the coevolution of software and hardware of Richard Dawkins, with software and hardware innovations pushing each other into a growing, and perhaps even inflationary, spiral, now without the need for the engine of prematurity. It would be as if prematurity had already fulfilled its role of initiating the process, which would henceforth be irreversible. I leave this question open, because these reasonings also involve many variants that need to be better defined, studied, and evaluated.

However, there is, in my theory, a strand that leads to a really startling result. Suppose my reasoning is right, and we have acquired human intelligence from the emergence of free memories, which have provided the existing algorithms with much more data and the emergence of new algorithms able to work with that data. There is no guarantee that human intelligence is stabilized, with an algorithm to control a database for language, a database for comparison, a database for images, a database for interconnecting multiple databases, and so on. Are there some human groups developing smarter brains and others developing less intelligent brains?

Theoretically and logically, this is not impossible. It is even probable because the phenomenon of differentiation occurs in every generation of living beings. Differentiation and adaptation are general rules of evolution. Will there be a cognitive regression for some humans and a progression for others? Reader, please note that I am not talking about racial or ethnic differences. I am talking about brain differences, ones that are apparently imperceptible and that could happen within the same races or ethnicities. In this way, a new group of smarter humans, independent of color or place of origin, could gradually be taking over the planet, to the detriment of another group that would be relegated to the background. This may be happening even with subtle changes in mating preferences.

Part III

Perfect fit

12

First list:
the initial behaviors

When I explained how human intelligence came about according to my theory in Part I, I drew up two lists of events that would have occurred between our separation from chimpanzees and bonobos and the first vestiges of an intelligence similar to that of modern humans appearing on the planet: one that began about 7 million years ago and ended 2.5 million years ago and another that began 2.5 million years ago and ended 50 thousand years ago. For that part of the explanation, they seemed adequate and sufficient to me because, at that moment, I did not want to present the proposal that the extraordinary eruption of Toba would have been the cause of the phenomena that led to the emergence of human intelligence because the purpose was to explain the logic of the theory, not the provocative phenomenon of pelvic narrowing. But the explanations could have continued without the necessary inclusion of Toba, as the second list shows, ranging from 2.5 million to 50 thousand years ago because, in principle, the theory does not depend on Toba.

But considering the natural imprecision of datings and interpretations, the connections both temporal and causal of the Toba Volcano eruption with my ideas seems well fitted and adjusted. Although I defend the Toba Theory, although I don't discard other explanations that may fit my Human Intelligence Emergence Theory, especially those that point to climatic occurrences on the African continent, such as that proposed by archaeologist Curtis W. Marean.

Thus, considering that the great eruption of the volcano in Sumatra caused pelvic narrowing, I chose to divide the second list of Part I into two new lists—one from 2.5 million to 75,000 years ago and another from 75,000 to 50,000 years ago; then, I propose that, in a general way, modern humans would have completed their formation process, according to the understanding of what is now considered to be human. The explanations made hereinafter concerning the events listed in the first and second chapters of Part I and now repeated in the second, third and fourth chapters of Part III are to be seen more as a complement to what has already been cited or proposed in Part I, instead of as something that has a new interpretation. I had chosen to explain some topics at that time because I found them to facilitate understanding, and I prefer to explain others now, for the same reason.

However, from now on, the arguments also have the sense of showing the viability of the Toba eruption as the causal factor of the emergence of human intelligence. In this sense, the difference between Part I and Part III can be summarized as follows: Part I demonstrates the logic of the Human Intelligence Emergence Theory but does not exactly point to the motivation of the initiation of the process; Part II considers the theory to be true, reexamines a few points, and explains others while arguing that the eruption of Toba is viable as the cause of the narrowing of the pelvis, a detonating phenomenon and feeder of the emergence of human intelligence, to my understanding. Truly, I tried to write all of Part I as if I did not know the contents of Part II and Part III. Nevertheless, some subjects are better clarified now, more for a didactic purpose.

It is worth noting that, in general, all the events and reasoning explained in this Part III somewhat support my theory with a different event having caused the pelvic narrowing in another period. Of course, there is an acceptable timeframe for my theory to be logical, which I imagine to be between 500,000 years ago—when we separated from the lineage that evolved into Neanderthals, and 60,000 years ago and when we have at least a space of 10,000 years to change from a hominid with chimpanzee-like cognition to a hominid with cognition like that of modern humans. Within this range, I firmly believe that the Human Intelligence Emergence Theory is possible to be explained.

I also add a new list, called the FIRST LIST, which covers an indefinite time of the past up to 7 million years, in order to comment on some human characteristics which, I imagine to have arisen before our separation from chimpanzees and bonobos, and which I consider important for the understanding of what modern humans really are.

Again, I note that these lists have more of a didactic character than an intention to represent a rigid schedule of events, even acknowledging that it is an attempt in that sense. And I note that the explanations for the initials after the events are the same ones already reported in Part I. This said, I now present four lists:

FIRST LIST (Part III)
Some date before 7 million years ago
- desire to hunt and to commit violence against other animals (ea);
- characteristic of kindness and friendship (ea);
- violence against individuals of the same species (ea);
- theft from both the same species and other animals (ea);
- replacement of claws by nails;
- loss of the tail.
End of the period: 7 million years ago

SECOND LIST (Part III)
Beginning of the period: 7 million years ago
- separation of the ancestor of humans from chimpanzees and bonobos;
- loss of long projection canines (ea);
- replacement of long projecting canines with stones (ea);
- increased use of bipedal walking;
- adoption of a fully biped locomotion;
- first extension of childhood;
- emergence of crying of babies (ea).
End of the period: 2.5 million years ago

THIRD LIST (Part III)
Beginning of the period: 2.5 million years ago
- large increase in brain size;

- appreciable improvement in balance in the bipedal stage (ea);
- increased accuracy in throwing objects (ea);
- enlargement of buttocks (ea);
- great increase in height;
- growth of the penis (ea);
- fabrication of stone tools;
- small increase in quality of stone tools.

End of the period: 75 thousand years ago

FOURTH LIST (Part III)

Beginning of the period: 75 thousand years ago: eruption of the Toba volcano

- small reduction in the opening of the pelvis (ea);
- small reduction in brain size (ea);
- decreased gestation period (ea);
- second prolongation of childhood (ea);
- change in nose shape, with nasal cavity pointing down (in);
- thinning of body hairs (in);
- appearance of a layer of fat (in);
- turning of the lips (in);
- control of respiratory movements (in);
- ability to also breathe through the mouth (in);
- laryngeal diminution (in);
- sweating as a way of controlling body temperature (in);
- fire control (ea);
- cooking of food (rw);
- sense of immortality (ea);
- idea of God (ea);
- female orgasm (ea);
- occult ovulation (ea);
- constant sexual receptivity (ea);
- menopause (ea);
- increase in lifespan (ea);
- monogamy (ea);
- mating preferences (ea);

- male and female adultery (ea);
- private sex (ea);
- male and female homosexuality (ea);
- African genetic diversity (ea);
- Modern humans migrate to Asia and Oceania (ea);
- Clear marks of a human intelligence are detected through the sudden appearance in archaeological records of spearheads, drawings, paintings, sculptures, statuettes and ornaments.

End of the period: 50 thousand years ago

A careful perusal will reveal that the first list in Part I is the same as the second list in Part III and that the second list in Part I is a combination of the third and fourth lists in Part III. In short, there are only two changes: the addition of the first list and the division of the second list into two. The addition of the first list, as I said earlier, is to explain some properties that arose prior to the separation of chimpanzees and bonobos; the division of the second list into two is to show more clearly the fit of the Toba explosion into my Human Intelligence Emergence Theory, with the division of events before and after Toba.

I could not deduce when the first characteristics that I mentioned as occurring before 7 million years appeared: the desire to hunt and violence against other animals; propensity to goodness and friendship; violence against individuals of the same species; and theft from same and other species. These aggressive qualities, which are invariably directed toward the need for animals to feed themselves, especially of the meat of other animals, have their origin in the times when all living beings were still in the oceans. Without going further, since 460 million years ago, when the common ancestor between humans and sharks and rays lived, these faculties have been rooted in our ancestors. Today, we have several animals that have lost their aggression toward other animals because they have become herbivores, but, even among them, disputes over food or females cause aggression and even death among the species itself; for example, as seen in buffalos, elks, and goats.

For evolution, eating is essential—attacking, killing, stealing, and making friends all stem from and revolve around food. Food, a daily obligation, is even more important than procreation. It is clear that the lack of one or the other ends the species, but it is also clear that a single sexual relationship is enough to guarantee the species, while feeding has to occur every day, and even at all hours, for some animals. Although procreation does not have as much urgency as food, animals also strive, kill, steal, and generate relationships to procreate. We modern humans, even with a privileged intelligence that has provided us with a sophisticated culture, seem not to differ so much from other animals in these aggressive characteristics (DIAMOND, Jared. *The Third Chimpanzee*, pp. 12 and 15). Robberies, theft, rapes, and murders are part of our history and of daily news around the world. Human intelligence, rather than diminishing the effects of these attributes, seems to have increased them and even perfected them. On two points we are absolute champions. First: no animal kills more animals of different species than we do. Second, no animal kills more animals of the same species than we do. We have created industries in order to kill other animals and feed on them. Everything is done professionally, creating them and fattening them up quickly for slaughter, in order to maximize profits. The disparity with other animals is so astronomically immense that it is an irrefutable singularity. It is even shameful, according to some ethical standards.

Nor are we pious with our fellows. But now, except for minor, not well-publicized exceptions we no longer feed on them, as we have done in the past, and as is the case of reciprocal cannibalism and rituals of torture, rape and sadism which still occur among pygmy tribes and other ethnic groups from Africa. It is a continent, moreover, that lives in a state of general civil war, which, in a few years, produced 3 million deaths and 2 million refugees, according to the Brazilian magazine *Veja*, edition no. 1,785, dated 01/2003.

However, do not think that I'm racially biased in citing the African problem. In terms of wars with an equal or worse history of absurdities, Europe is the cradle of white civilizations, where, just to mention more recent cases, two World Wars occurred; during these, the most scabrous cases of murders and rapes of innocent men, women, and children were recorded in hundreds of village invasions. Democratizing the horrors, all these facts are repeated in Asia, the Middle East, Oceania, and the three Americas. All over the planet, wherever modern humans live, this circus of atrocities is present. Unfortunately, there are no exceptions. In Brazil, a person is murdered every

9 minutes and 48 seconds. In the USA, every 34 minutes; in Japan, every 13 hours and 22 minutes, and in Canada every 14 hours and 22 minutes. In the rest of the world, unfortunately, the picture is the same, with a tendency approaching the rate of murder in Brazil, especially in the poorer countries.

Some say that humanity is lost, that we are born good and the world turns us into bad. Nothing is more misleading. In fact, mankind has been much worse! Culture has improved and has applied rules to humanity, almost always with threats of punishments. However, this improvement has been very small. Aggression, killing, and stealing from fellow creatures and other animals, therefore, are attributes that we acquired millions of years ago and which remain present as if to remind us from where we came.

Proof of this is everywhere. Doors, gates, electric fences, padlocks, and keys exist for only one purpose: to prevent humans from being attacked, killed, or robbed by other humans. Police officers, security guards, lawyers, and judges were created with one purpose: to prevent humans from being beaten, killed, or robbed by other humans. In neither case do these measures eliminate the behavior. In the first case, our fellow men always discover new ways of overcoming any security system. In the second, our fellow men, with so much power in their hands, are in charge of their own instincts. We are all potential hunters, murderers, and thieves. We just need a reason for the instinct to surface. What controls us is culture, through rules and punishments. Dictators, presidents, kings, ministers, senators, deputies, secretaries, councilors, pastors, priests, judges, lawyers, doctors, engineers, economists, businessmen, and the general population steal, attack, and kill whenever they know they can do these without consequence.

But these skills would not have the evolutionary efficiency they possess without another, say, more general and diversified characteristic: the ability to lie. The principle is the same: living beings need to feed on other living beings, and, for this, they need to hunt. Usually one does not hunt without deceit. One also does not usually kill or steal without deceit. Feeding on other animals requires a good masking ability, as occurs with several species, such as our domestic cat, which lowers itself when hunting for mice, or some strange fish, which disguise themselves as stones in order to swiftly swallow fish that may inadvertently approach.

149

However, we also have the instincts of kindness and friendship, which, in the same way, I think, emerged well before we separated from chimpanzees and bonobos. At a time when the approximation among similar beings became an evolutionary advantage, the foundations of the instinct of friendship were probably laid. I believe that what we call love is a mixture of friendship and procreation instincts.

Nail monkeys, in my home state of Piauí, Brazil; chimpanzees and gorillas in Africa; and several other animals around the world live in groups, helping each other. If they do so, it is because there is an evolutionary advantage. This is the general rule of evolution. Friendship, as we humans understand, is the basis of cohabitation and coexistence. It is not easy to understand the coexistence of instincts as diverse as killing and stealing and kindness and friendship. But that's the way living beings work. A male dog that is sniffing and licking a female dog at one point is the same one that advances to kill it five minutes later, fighting over food. The chimpanzee that affectionately picks mites off a friend can join this friend to kill the leader of the group where he lives so that, together, they take the lead. The first thing a lion does after defeating a pride leader is kill all the cubs to ensure that only his own offspring live thereafter. The lionesses, who watch their cubs being killed, mate with the killer to have other cubs. But all this, strange to the precepts of human cultures, has a terrible adaptive logic: to pass genetic information to the greatest and most qualified number of descendants. Because, in that way, they will have more descendants with the same qualities, and so on.

I will now carry out an analysis of these faculties, one uncommitted to human cultures and their concepts. Whenever we think of violence, robbery, theft, and murder, we think of evil in the cultural sense. And whenever we think of kindness and friendship, we think of feelings of altruism. But there is another more realistic and less romantic view that these two behaviors have less noble, hidden objectives of interest in the survival of those who demonstrate them. To be as direct as possible, by this judgment, one who practices kindness and friendship would simply be putting himself in a situation of deserving reciprocity in the future. Kindness and friendship would work as a kind of savings for worse days to come. The one who showed himself to be a good friend would be doing nothing more than increasing his potential for survival. He would thus be a good friend for his own benefit.

What we understand to be love would also have its practical motivations, among them, of course, procreation and child rearing. To summarize: passing on one's genetics is the general rule of species evolution.

However, if this takes away all the connotation of feelings that culture has established for the occurrences identified with altruism as goodness and friendship, it also performs a similar function for the skills identified with cruelty, such as theft, murder, and rape. Everything would also be done with the same intent of survival. If we could replace all these behavioral traits with a single sentiment intelligible to today's human culture, it would be selfishness. This idea is similar, I imagine, to what Richard Dawkins suggests, in his book The Selfish Gene. It's just that Dawkins centers on the figure of the gene, and I, in this moment of reasoning, center on the figure of the human species. In other words, on the individuals who comprise the species.

If the behavioral properties that I cite emerged hundreds of millions of years ago, the two physical properties that I considered relevant occurred more recently, if I may in this way express myself: the replacement of claws with nails, about 60 million years ago, and the loss of the tail, 25 million years ago. As I said in the second chapter, the first seems to be a consequence of the advantage of having equipment more suitable for digging than for hunting, and the second, because of the lack of the tail's utility at some point in evolution.

Both had important implications for hominid development. The lack of claws limited the power of hunting and must have pointed, in some era, in the direction of vegetal feeding. Tail loss, however, directly affected balance and displacement, perhaps even requiring a further increase in brain size to provide the organism with a network of more efficient neurons to control the legs.

I think that, after explaining these attributes, the reader should imagine a very small chimpanzee-like animal swinging through the tree branches of the dense forests that probably existed in Africa 7 million years ago. One of these animals, for some reason, likely climatic or ecological, gave rise to two species: the ancestor of modern humans and the ancestor of chimpanzees and bonobos.

13

Second list: bipedalism

In order to remind the reader of my proposals, I will now briefly review what I explained in Part I about the events that I relate in the second and third lists, with small explanations that were not previously mentioned, with more relevance in order to strengthen the idea that the narrowing of the pelvis would have been triggered by the explosion of the Toba volcano. The reader should refer to the second list in the previous chapter.

Summarizing my thoughts on this period, I recall that I proposed a new theory for the emergence of bipedalism, considering as the main cause the replacement of long projection canines with stones as hunting weapons, which would have been used as a kind of hammer or throwing projectile, which made our ancestors perfect bipedal walking, so that they could carry the stones with their hands, while moving quickly using only their feet. I also imagined that in this period there occurred an extension in childhood, to allow the learning of that complicated locomotion and the technique of throwing stones, which is much more complex than it seems. Finally, I presented an explanation for the crying of babies, which in my view corroborates the causes of bipedalism, the need for mothers to carry their children on long walks. I think, then, that the loud crying of our babies is more a protest of not being in the mother's lap, rather than a warning of some physical suffering. With these developments, the period ends. At this point, I think that a new reading **Bipedalism** (Chapter 3) and **The cry of human babies** (Chapter 4) in

Part I, would make the understanding of the explanations to be followed in chapter 14 more comfortable.

But, how would our ancestors have been at the end of these events? I believe they would have been animals with many of the physical characteristics of chimpanzees and modern humans: with a brain around 420 cm3, a height of about 1.2 m, practically without long projection canines, with hair like chimpanzees, and a reasonable agility in using stones as a hunting weapon. They would have almost a completely bipedal life, but they still would not have completely lost their ability to move around in the trees. Ah, yes. And with the babies crying unbearably. Anyone who raised children as I did knows what I mean.

14

Third list:
the growth of the brain

If the previous period was notable for the appearance of bipedal walking, the period of the third list is notable for brain growth. Here, a new reading of Chapter 6, **The extraordinary growth of the brain**, seems appropriate to me, because it shows my position contrary to the general understanding of science that brain growth would have had a direct influence on the rise of the cognition of modern humans. But, in a way, I owe an explanation for the extraordinary increase in brain size without the need to be tied to the increase in human cognition. The reader can follow the third list in chapter 12.

Ever since I developed my theories, I have been thinking of finding a reason for the brain to have increased so much without necessarily having a direct relationship with an increase in intelligence. I know it's very hard to fight the idea that we're smart because the brain has grown. After all, only our brain has grown so much, and we alone have the differentiated intelligence of modern humans. However, doubts are already beginning to arise about this, because more recent research shows that some birds, such as crows in New Caledonia, have a proportionally larger brain relative to the rest of their body than most animals, but this only seems to be associated with an amazing ability to build and use one type of tool.

But I will develop a different reasoning. Did not the coincidence of these two facts, the extraordinary growth of the brain and the emergence of human intelligence, provoke a myopia that confused our view of what actually happened? Brains have grown, it's true. However, there nothing indicates that humans were increasing their intelligence before 50,000 years ago. If I am right, the brain has grown for another reason. And here again comes my project of sincerity. I certainly do not want my explanations of brain growth to be interpreted as stubbornness, where I only share facts that fit my theories. That's not it. It is simply my search for a solution, since I don't agree with what is currently accepted.

I consider two characteristics logically acquired in the period in which the brain grew as my favorites to explain this phenomenon: balance and the throwing of objects. I know they don't appear as charming as the rise of new cognition, but they require complex cognitive ability much more than you think. It's worth noting that, although formidable, the growth of the brain took about 2 million years to cause an increase of 400 cm3 to 1400 cm3, when our ancestor humans grew from 1.2 meters to 1.8 meters in height, which may explain part of the increase in brain volume, since it is a general rule that, when an animal grows, its whole body grows proportionally. Dawkins, in his book *The Ancestor's Tale* (pages 104 to 114), shows a way of interpreting the relationship of brain size to animal size using logarithms, which indicates that the hominid brain has grown out of the ordinary. Unfortunately, I didn't understand the explanation. What I understand is that, if the height of a solid increases twofold along with all other linear measures, its volume increases by eight. Using this criterion, if the height of hominids went from 1.20 meters to 1.80 meters, the volume of their brain, that was 400 cm3, was expected to grow to 1,350 cm3, that is to say, a value very close to the result found, of 1400 cm3, according to what I have found in several books and texts. In Figure 7, Chapter 6, I present a graph in which the reader can observe the relationship between hominid height and brain volume during the course of human evolution over the last 7 million years.

In any case, since it is consistently accepted that the increase in brain volume was the main cause of the increase in human cognition, in this book, whenever I deal with the subject, I will not fail to cite this increase as extraordinary, although I don't agree with the exaggerated emphasis given to this cause and effect relationship.

Moreover, I find the way bipedalism is usually treated to be very shallow. To be rigorous, hominids are actually the only biped beings on the planet,

155

even when compared with extinct animals such as dinosaurs. Dinosaurs were never perfect bipeds, because they used their tails to walk or rest. And what about birds, which are their descendants, according to the latest scientific explanations? They aren't bipeds, either. Birds use their wings and tails for both locomotion and balance. Both the ostrich and the emu, when they run, use their wings to hold themselves up and not fall, as do chickens. Kangaroos? They also use their tails for support. It seems that bipedalism, precisely defined, is indeed a human singularity. It is a singularity that seems to have nothing to do with the other singularity, human intelligence, because we were bipeds for at least three million years without possessing human-like intelligence, so much so that I situate the emergence of bipedalism as the most important event of the second list.

Given that the events are all intertwined, and the lists have only a didactic and organizational purpose, I think that when bipedalism was established as the main means of hominid locomotion, there was a need for a larger brain to control their balance, which was now concentrated in the legs, with little help from the arms moving like pendulums in faster movements. I imagine that the more hominids adopted biped behavior, the more they needed to have agility in their new locomotion, and the more the brain grew to perform the complex computational procedures required to balance an erect body on only two legs. When we are walking, we are left with both feet on the ground for very little time. We are practically balanced on one foot or the other, and always on a minimal part of that foot. When we run, maintaining balance is even more demanding, because, with each step, we stand in the air without any support. Our way of getting around seems to be very complicated. Only when we are approximately one year old do we begin to take the first wobbly, stumbling, unbalanced steps.

Would this be enough to explain the exaggerated growth of the brain? Maybe not, but if you rely on another force pushing brain growth in the same direction, like the throwing of objects, it might be. I have previously explained that our ancestor hominids replaced their long projecting canines with using stones as hunting weapons. I think that stones were not only used as fixed hand weapons. I think, and I see much logic in this, that they were also used as projectiles. Even by the number of supposed hand axes found, the use of which remains questionable (KLEIN, Richard G., BLAKE, Edgar.

The Dawn of Human Culture, p. 91). William Calvin, an American theoretical neurophysiologist, proposed that ballistic movements, such as those used to launch projectiles against distant targets, require special computational operations of the nervous tissue (DAWKINS, Richard, *Unweaving the Rainbow*, p. 380). I imagine that this requirement may also help explain brain size increase without the obligatory emergence of human intelligence. The need to throw stones accurately as a way to hunt small land animals, and even birds, may have required a complexity of calculations in the brain, which forced an increase in brain volume in order to establish the necessary neurological connections to command the muscles so that they could perform the movements with necessary accuracy. In this way, the first three events would be directly linked: **the considerable improvement in balance in the bipedal gait, the improvement in the precision in the throwing of objects, and the increase in brain size.**

But bipedalism is related to other human attributes. By observing animals, especially those most closely connected to us such as great primates and monkeys, it's clear that only we have voluminous buttocks. Like every singularity, it deserves an explanation, or at least an attempt at an explanation. I propose that the buttocks have arisen to protect the anus from parasites when we descended from trees and assumed bipedal locomotion, since we no longer had a tail to frighten them. Perhaps this even happened in the previous period, during the beginning of the change in locomotion. In any case, if the buttocks had increased by that time, I think this also caused an increase in the size of our ancestors' penises, an occurrence that many authors still acknowledge does not have an acceptable explanation. Among the great primates, modern humans are those with the largest penis (on average 13 cm). Gorillas have the smallest (on average 4 cm), and chimpanzees and bonobos, are intermediate (on average, 8 cm). The logic is simple: an increase in the volume of the buttocks creates an obstacle to penetration of the penis into the vagina, since it is thought that hominids, like the great majority of animals, had sex with the male positioned behind the female. To solve the problem of bulky buttocks, nothing was simpler than having a larger penis to overcome the obstacle of the new distance, transporting the sperm to the vagina with greater security.

But I also see another reason for the increase in hominid penis size.

157

With the adoption of the bipedal posture and the consequent change in the position of the spine, the female's vagina shifted downward, also hindering penile penetration and also favoring an increase in length to compensate for the increase in distance. These are two propositions that converge rather than oppose each other. Independently, one can survive without the slightest need for the other to be true.

In a chapter devoted to the increase in brain size, I ended up talking more about bipedalism. My reasoning led to this, since many changes that occurred with our ancestor hominids were due to this new way of getting around. Even bipedalism, having established itself in the second list period, from 7 million to 2.5 million years ago, was caused by many transformations, including brain growth itself, and ended up on the third list, from 2.5 million to 75 thousand years ago.

The great increase in height can only be credited to the advantage of confronting larger animals, much more numerous on land than in the trees, both in hunting and in the fight for food, but above all in the struggle for defense. Fossil records show that by this time, humans were being decimated by large predators. To grow was therefore to increase the possibilities of defense, especially if they lived in a group. The making of stone tools and the small increase in the quality of these tools are well explained in Chapter 6, **The extraordinary growth of the brain**, when I try to demonstrate that these facts have nothing directly to do with increasing human cognition.

And how would our ancestors have lived in the extreme south of Africa, at the end of the third list, before the explosion of the Toba volcano? The hominid that fits my proposal is undoubtedly *Homo sapiens idaltu*, represented by the fossils of Herto, in Ethiopia: three well-preserved skulls, two adults and one child. The fact that they were found in Ethiopia, far from the extreme south of Africa, is of no importance to my theory, because they were dated 160,000 years ago, that is, long before 75,000 years ago. This period explains the displacement of the species, or a descendant of it, to the south of Africa. What matters is that there were hominids who lived 160,000 years ago with brains the size of modern human brains, and even slightly larger.

Thus, I suggest that, 75,000 years ago there lived in the extreme south of Africa a biped hominid about 6 feet tall, with a brain a little larger than ours, limited ability to hang on branches, a coat similar to that of a chimpanzee, and a level of cognition very similar to that of most animals. These last two properties deserve a little explanation, because they confront the understanding of science today. Whenever I come across illustrations of hominids from that time, they are portrayed as almost hairless. I don't see the reason for this. Most of the animals living in Africa had thick hair, which can be deduced as an advantageous adaptation. The exceptions are those who their lives partially in water, such as hippos, rhinos, and elephants. As for having a cognition equivalent to that of other animals, this is a basic idea of my theory, and it is based on the fact that no clear and reliable traces of human intelligence have been found that date back to more than 40,000 years ago. What these hominids did as an extraordinary thing, and which, I acknowledge, seems to be an activity which demanded a better developed cognition, was to chip stones and use them for hunting and carving up animals. But, as I have explained, several animals perform similar activities and don't possess a cognition that is even close to that of modern humans.

15

Fourth list:
human intelligence

The first event of this period, from 75,000 to 50,000 years ago, was undoubtedly **a small reduction in pelvic aperture**, caused by the urgent need to seek food at sea and to perform endurance hunting, as I explained in Part II, Toba. Now, I will address a suggested reading for a better understanding of the subjects. Therefore, I consider the events directly related to the appearance of human intelligence to be well explained: **the small reduction in pelvic aperture, the small reduction in brain size, the reduction of the gestation period (prematurity), and the second prolongation of childhood.** All these flow and influence an inflationary process of the emergence of free memories in hominid brains, which, according to my theory, caused the emergence of human intelligence and, consequently, the origin of humanity.

At this point, I consider it important to remember that, before settling on sea hunting as the main effect of the extreme lack of food in South Africa and therefore as the main cause of pelvic narrowing, I had great sympathy for the theory of endurance racing and hunting and had even begun to study this possible result, observing documentaries of African tribes hunting antelope—especially the San people (a Khoisan subgroup and also called bushmen)—and the extraordinary endurance in marathon races demonstrated by the people of Kenya.

Today, although I consider sea hunting to be the main cause of pelvic narrowing, the idea that endurance running has also been an essential necessity in the extreme south of Africa after the Toba volcano erupted is not entirely out of the question. This theory finds support in the physical properties of modern humans, such as legs with a molded muscle for prolonged exertion; feet and ankles strong enough to absorb and withstand repetitive impacts; sweat glands that efficiently cool the body; and an extraordinary respiratory system, which allows long runs (including breathing through the mouth or nose), with a variety of speeds, such as those used by soccer players during a game lasting 1 hour and 30 minutes divided into 2 periods of 45 minutes, with rest of 15 minutes between one period and another. On the other hand, antelopes, victims of this hunting technique in spite of being very fast— reaching speeds up to of 100 km/h—release heat through respiration, which, after a few hours, causes their body temperature to reach levels that kill them or force them to stop running. In a resistance hunt, between the hunters of the San people, whom I mentioned briefly in the last paragraph and who still live today in the Kalahari Desert in southern Africa, and the antelope, the end is predictable: many hours after the pursuit begins, the antelope stops, exhausted, and the hunter approaches with a spear to kill it.

In any case, the two events, sea-hunting and endurance hunting, are reciprocally compatible, with the potential do aid each other for explaining a decrease in pelvic aperture. This potential of reciprocal compatibility does not seem to be so rare in evolution. Two examples of my reasoning illustrate this well: the explanations of bipedalism and the exaggerated growth of the brain. As the main cause of the first, I propose the replacement of long projection canines with stones, but I do not reject other explanations, such as the change to a savanna environment, the need to carry babies, the lower exposure of the body to the sun to regulate temperature, and other similar factors. As crucial motifs of the second, I point to equilibrium in the bipedal gait and the skill of throwing stones, but I do not completely rule out the possibility that a small increase in cognition for the purpose of making stone tools has also had an influence, even if it was small.

Continuing down the Fourth List, we now have a series of events that Elaine Morgan mentions in her book, *The Aquatic Ape Hypothesis*, as consequences of a semi-aquatic phase that hominids would have experienced 5 million years ago before returning to a predominantly terrestrial lifestyle, and which I think occurred, at least in part, 75,000 years ago, when the life of Homo sapiens who escaped the Toba catastrophe in the extreme south of Africa depended above all on sea hunting and endurance racing: **a change in nose shape**, with nostrils pointing downwards, preventing drowning during sea hunting; **thinning of the body's hairs**, facilitating the hunt by reducing friction with the water, and giving the false impression that we are naked apes; the **emergence of a layer of fat**, as protection from the low temperatures of the South African seas; the **curl of the lips** to contribute to the blocking the nostrils during swimming; the **acquisition of the ability to control respiratory movements**, aiding in the complex speech of human languages; **the ability to also breathe through the mouth**, assisting breath control while swimming and in endurance running; **sweat as a way to control temperature**, to refresh the body during endurance hunting; and a **descended larynx**, also corroborating in the emission of speech sounds.

The most difficult parts of the head to imagine evolving are the nose and ears precisely because rather than bones they have cartilage,, which is fossilized with difficulty. Is cartilage fossilization possible? I think so. Under special circumstances, of course, for to date no reliable fossil record of either nose or human cartilage has been found. The most perfect external-looking fossil I have located in my research was discovered by climbers in the Italian Alps, 3,200 meters above sea level, and dated 5,000 years ago. Called Otzi because it was found in the Otztal Massif, it was analyzed by numerous scientists from various countries who attested that he was not bald and had a beard, that he had been ill three times before his death, that he died in the spring, and that he fed on different types of plants and animals. In addition, he wore shoes made from the sewn animal skin and carried a copper ax with a wooden handle and a bearskin cap. But it was not possible to tell the shape of his upper lip, his nose or his ears because, it is supposed, the pressure of the ice had destroyed that evidence. But ice does not seem to be exactly the cause because fossils that actually show the size of the hominid nose over the last 7 millions of years have not been found anywhere: not in the blistering heat of Africa, in the cold of Europe, or in the varied climates of Asia. As far as my research can tell, the human nose is really a mystery that I will attempt to explain.

The change in nose shape seems to relate only to sea hunting, although it leaves a suspicion that it may have some relation to endurance racing. Chimpanzees, gorillas, orangutans, gibbons, and most monkeys have noses with nostrils directed horizontally forward, in a very dangerous position for those who need to hunt aquatically, since it facilitates, or better, forces the entrance of water, suffocating the animal and forcing it to return to land or die from drowning. The exception in monkeys, as far as I know, is unique: the Proboscis monkey, which lives in a partially aquatic environment, on the coast of Borneo in South Asia (MORGAN, Elaine. *The Aquatic Ape Hypothesis,* pp. 132 and 133). Therefore, it is a general rule, among apes and monkeys, that the nose is flattened, with the nostrils directed forward.

But this does not occur with modern humans, who have preponderant noses with nostrils directed downward, in a position that makes it easier to get into the water without drowning, such as by diving, when its shape acts as the prow of a ship, throwing water sideways and away from the nostrils. The idea I espouse is very similar to that of Elaine Morgan, Max Westenhofer, and Alister Hardy: humans, about 5 million years ago, had a semi-aquatic lifestyle. But I think this happened 5 million years later, and they did not exactly have a semi-aquatic life, but instead had an absolute need to acquire food found in rivers and in the ocean. For me, those humans in the extreme south of Africa who escaped the Toba catastrophe, by the absolute and urgent need to feed themselves on fish and other seafood, had to swim with great efficiency, which required a narrowing of the pelvis. In this way, I do not suggest exactly that they lived in both fluvial and maritime environments but that they lived on the land while decisively needing to hunt in the water, both in rivers and the sea.

Following the order presented, let us examine other aspects of the supposed entry of those humans into the sea. It is almost a general rule that all animals living in Africa had thick hair, such as lions, monkeys, and zebras. The hair would surely have provided some, or even several, advantages. Our close relatives—and DNA tests do not let us lie—chimpanzees, bonobos, gorillas, orangutans, and gibbons have thick hair all over their bodies. We do not. We are naked. In fact, we are not exactly naked. We even have more hair than chimpanzees do, only that hair is much thinner. This indicates that hair thinning occurred. What could be the explanation for this?

In my opinion, in the context of my reasoning, the reason seems simple.

163

No hairy animal can swim efficiently. Hair increases the body's friction with water, hindering movements and restricting agility. Almost no aquatic animal possesses hair like a chimpanzee, for example. In order to not overly extend this debate, suffice it to say that, in the Olympics, swimmers shave their body hair for better results. It should also be noted that in Africa, the only animals that lost their hair were those who had to live, even partially, in the water: hippos, rhinos and elephants. The latter, for those who do not know, are excellent swimmers. There have been reports of groups of elephants swimming for hours at sea from one island to another (MORGAN, Elaine. *The Aquatic Ape Hypothesis*, p. 70).

Fat, on the other hand, seems to be a predicate of two types of animals: hibernating and aquatic. However, I could not establish a logical connection between fat and aquatic life. Perhaps, something related to temperature control, such as conserving heat, could be a possible connection (MORGAN, Elaine. *The Aquatic Ape Hypothesis*, p. 76). In any case, I must mention this characteristic that is now becoming a global problem for modern humans, given the abundance of excessively caloric food now available to a large part of humanity.

Now **the twisting of human lips** is really a singularity. I find it strange that science hasn't studied and described this aspect more carefully. I've only seen it quoted superficially by Elaine Morgan. I think it deserves a more dedicated study. There is no primate or monkey that has lips like those of humans. They are twisted and have a central mark that fits perfectly below the lower part of the nose. If you, the reader, can lift your lips, you will see clearly that they can clog your nostrils, preventing water from entering if you dive. I know that it is not a determining fact, but it is something significant to guide my theory in the direction of a need to hunt at sea. Maybe it's even more important than I'm trying to prove right now. At any rate, I mention it here as another indication that modern humans have had a semi-aquatic life, or, as I prefer to emphasize, a terrestrial life in dire need of aquatic hunting.

Control over breathing movements is another faculty that most humans think all animals have. But it does not seem to be so. The rule, at least, is to not have this control, especially in the case of the animals that are evolutionarily closer to us. Dogs, for example, when panting, breath through the nose and exhale through the mouth. I don't know if, because of the demand of the hunt in the seas of the extreme south of Africa, we have a way of breathing that is different from that of most animals: we put inhale and exhale through both the nose **and the mouth.** The reader, quite rightly, may think that this is nonsense, just an evolutionary adjustment like so many others. But, besides being of great importance for efficiency in both hunting and endurance running, it may have been the characteristic that allowed for the emergence of language as we know it: we modern humans can talk and breathe at the same time. It may not appear so, but this is very important for our living together. The emergence of human intelligence required a complex, diversified, rapid, and efficient mode of communication, which in turn required a different breathing technique that would allow for maximum conversation without the interruption of basic vital functions such as breathing and even eating. Although it may not polite to do so, we modern humans can continue talking for a long time, even while eating! The expression "business dinner" does exist, whether it involves just talking about doing business or just eating if you can get dinner. Human language is, as I have said before, the fruit of a cognitive system capable of processing it, and it has also been developed using the differentiated faculty of sound emanating from human beings as a function of the demands of water hunting and endurance running. However, I admit that in this process, sound and respiratory specificities have undergone some modification to suit the needs of the complex way of communicating of humans.

Elaine Morgan also highlights **descent of the larynx.** I do not believe that this provided a chance for hominids to develop a language like that of modern humans. It is more feasible that hominids developed an intelligence that needed more complex communication, and they took advantage of what they had around them, such, for example a descended larynx, to use in their way of communicating. If the descent had not happened, something else would have been used. We could have even had a language without sound, or one with very few sounds.

I remember watching a documentary about a group of deaf-mute boys and girls who were taken to a distant school, if I remember correctly, in Central America, where they developed an extraordinary language of gestures that no one could properly decipher. They "spoke" at incredible speeds. It seems to be the same fact that Matt Ridley quotes in his book *Nature via Nurture*

(p. 214). So I say: the difference started in the brain. Before, we were like any other animal. Toba erupted, the pelvis narrowed, we were being born prematurely, babies were born with free memories, we were becoming smart, a more complex language developed, and a waterfall of events and situations resulted in the being that we are today. Particularly, I think we must urgently study ourselves with a humbler and less arrogant view, without absurd presumptions that we are the center of a universe that was created to serve us. The understanding of who we are, associated with what we have of good, and the recognition of what we have of bad, could change the destiny of humanity.

Finally, I believe that **perspiration as a way of controlling body temperature**, also addressed by Morgan, is more linked to endurance hunting. I think the need for this type of hunting may have been the reason for the emergence of this quality. Thus, I suggest that sweat is a response to hair thinning in the face of the problem of overheating, the result of endurance hunting. But there are many theories in this regard; among them, one says we lost our hair and we began to sweat because we went to live in the savannahs during a very hot time. I don't see much sense in this, because, in general, the animals that live on the African savannas have not lost their hair nor do they sweat. The rule in the savannas seems to be to have hair: chimpanzees, lions, zebras, antelopes, monkeys, wild boars, and gorillas all have thick body hair. Exceptions are animals with some connection to water—such as hippos, rhinoceroses, and elephants—or with air—such as birds, which have feathers. I must admit that I am still not satisfied with my conclusions on this subject. I lack more information, data and research that can lead me to a better understanding. Who knows—in a future book I may formulate a more convincing idea.

Meanwhile, there are other human faculties that indicate some relation between the evolution of modern humans and a partially aquatic life. I will not dwell on all of them because I am not entirely sure of various related matters and also because to do so lies outside this book's primary purpose. But for the sake of an example, just to note that even newly published facts point to this connection, I read in 2013 news of an experiment that suggests that our hands are wrinkled when wet to facilitate the handling of smooth objects in water. This fact points to early hominids hunting crustaceans and fish more easily on the beaches of South Africa, as I propose. In any case, if this book is taken into consideration, someone more prepared than I am can clarify points about which unfortunately I do not have the technical knowledge to explain at the moment.

With so many instances pointing toward the same fact, I find it reasonably possible that modern humans, due to the climatic phenomenon triggered by the Toba explosion, lived a partially aquatic life some 75,000 years ago. As the climate normalized around 60,000 years ago, they had the chance to spread across the African continent and the rest of the world, now equipped with intelligence capable of qualifying them to live in the most varied ecological systems.

Next, in Part IV, **The past explaining the present**, I begin to clarify events starting with **fire control (ea)** on the Fourth List. Even with the conviction that these events occurred simultaneously, almost certainly between 75,000 and 50,000 years ago, the type of division I use allows me to relate each one in its space, with its title, facilitating the understanding of its particularities and enabling a better understanding of the whole process. The reader, therefore, will better understand the arguments accompanying the fourth list, in chapter 12, starting with fire control (ea).

So far, I can say that I have been dealing more with physical changes, such as the reduction the pelvic aperture, the small reduction in brain size, and the decrease in the gestation period. Starting with the next chapter, I will deal more with behaviors: fire control, cooking food, the sense of immortality, and the idea of God. I have also combined a few events on the same topic, such as controlling fire with the cooking food and the sense of immortality with the idea of God because I have noticed that the explanation of one has many connections with that of the other. I close Part IV with the different chapter of the context, 27, **The possibility of proof**, where I outline an experiment to test whether prematurity, occurring with a species genetically close to modern humans, in a manner similar to what I say happened to a group of hominids in the extreme south of Africa, may cause an increase in their cognitive ability.

Part IV

The past
explaining the present

16

Controlling fire and cooking food

To explain controlling fire and cooking food, I turn to Richard Wrangham, who in his book *Catching Fire - How Cooking Made Us Human* demonstrated how cooked foods are far more nutritious than raw foods. However, I do not agree with him when he says that we may have started cooking before becoming intelligent and that human intelligence as a consequence of cooked food. Nonetheless, I accept his position on the nutritional value of cooked foods, especially when he indicates their need to maintain a brain with special human cognition. I admire and agree with his logic of thought. I simply do not think it happened in that order and at that time. I therefore also accept that boiled foods really were important in sustaining the new intelligent and "energy greedy" brains, not 2 million years ago, but rather 75,000 years ago when hominids were already developing a cognition similar to that of modern humans.

Thus, with a similar reasoning to that which I applied to language, when I suggested that intelligence came before the complex modern human language, I also argue that intelligence came before the ability to maintain a burning fire, absolutely necessary for the cooking of food.

With due apologies to Wrangham, whom I thank for the knowledge I acquired from his book, I will explain my disagreement: I consider the operation of keeping a fire burning to be so complex that it could only have been carried out by a being with extraordinary cognition like that of modern humans. Making a nest as do birds, however intricate it may seem, is a repetitive activity. Building dams as do beavers involves some complications, but these are always solved by using more or less material. Controlling a campfire is completely different. It requires complex decisions at every turn, involving countless variables. Also, each fire is unique. It has peculiarities that require rapid and decisive reasoning, both to keep it lit and to prevent the fire from spreading through the environment, as well as choosing the appropriate tree for its wood, the humidity of the wood, the humidity of the weather, the wind, and various factors. Thus, I believe it is impossible for an animal to first learn to cook and to then become intelligent. In my opinion, it is exactly the opposite: first came the cognition, then the cooking. Of course, there is no watershed in these processes. Cognition arose slowly, and I cannot tell when this cognition gave humans the control of fire, or, more precisely, the beginning of fire control and consequently the cooking of food. What I take for granted is that fire control only occurred after the development of human intelligence was at least already well underway. The inflation of which Dawkins speaks could be possible, and one thing pulls the other: free memories provide intelligence, intelligence provides cooking, and cooking provides the energy to increase intelligence. Thus, between 75,000 and 50,000 years ago, when exceptional human intelligence was established, our ancestors in the extreme south of Africa began to control fire and to cook food, favoring the sustenance of the emerging organ that required so much energy: the new human brain.

The sequence of events, I repeat, in order to avoid misunderstanding of an inflexibility that does not apply to the explanations presented in this book, must be considered as didactic. Although, of course, I try to place events in chronological order whenever possible. I think the facts I have mentioned so far in this chapter actually occurred as early as the first millennium after Toba's eruption, but I cannot tell when they ended. I can only say that those hominids who escaped in the extreme south of Africa became modern humans at least 50,000 years ago because this is the date now accepted by science of the arrival of Aborigines in Oceania. To state the opposite, as I have already mentioned, would be the same as saying that Aborigines are not humans. Then, I would be considered a worst kind of racist, and I do not want any confusion in this regard. Besides, honestly, I do not have the technical conditions to think about it, since I have never found a book specializing in human fossils found in Oceania. In any case, the precise date of the birth of the first modern human is not very important for my theory, as long as it is compatible with the arguments I present in this book. New discoveries and new techniques of more precise and reliable dating will certainly settle many doubts in the near future.

17

The sense of immortality
and the idea of God

In this chapter, I will address two controversial and delicate arguments that may cause outrage in some people: the sense of immortality and the idea of God. I do not want you to think that I am mocking religion. In the introduction, I already stated my position on this subject, which I repeat now to avoid all confusion. I am an atheist in health and religious in disease. In other, better words, when I'm in trouble, I appeal to religion; when everything is going well, my doubts and logical reasoning return. In short: I am human. I am not ashamed of this dubious situation, because my conclusions show that a religious sense is innate to human beings. A successful evolutionary adaptation has given us this characteristic.

When I am religious, I consider myself to be a Catholic, for that was the religion that my mother and grandparents taught me. At certain times in life, clinging to a religion is very comforting and advantageous.

When I refer to the emergence of the sense of immortality, I'm referring to the emergence of a specificity that causes all humans to tend to believe in the existence of an afterlife, as all religious beliefs basically claim. When I speak of the idea of God or of gods, I am not referring to the idea of the God of Christians, of the Jews, of the Philistines, of the Muslims, of the East Asians, of the ancient Egyptians, of the ancient Romans, the ancient Greeks, the American Indians, the Brazilian Indians, the Australian Aborigines, or any other Gods from various religions around the world. I am referring to the emergence of an idea that one or more deities somehow protect modern humans who fulfill certain formalities, usually linked to submission, contribution, sacrifice, and social norms.

If there really is a God or gods, I cannot give an opinion in the context of this book because this is a matter of faith; as I have already stated, I am dealing with science here, and science requires experience, proof or reasoning. Reason indicates to me that, along with the emergence of human intelligence, the consequence algorithm has emerged. What is the consequence algorithm? Algorithm allows modern humans to calculate what will happen in the future by observing present or past facts and events. It is the algorithm that allows you to plan. Other animals hunt without considering the danger. Humans do not. When they became intelligent, they came to understand the risk of hunting. They might not come return alive. From their young days, they would saw their parents going to hunt and sometimes not returning. Or, they would return ravaged by the beasts they faced. Their thoughts were probably haunted by the unpredictability of the world in which they lived. They needed to rely on something to mitigate their anguish. Clearly, behavior that gave them hope would be advantageous. So, when they came to have a conscience and understand that their lives depended on so many random facts, there was an extreme need to put fate in the hands of a higher being—something that could protect them through established behavior, such as making an offering and receiving a grace. Perhaps the sense of immortality began with the comparison of sleeping and waking up with dying and resuscitating in another life. The fact that humans observed sleeping and waking humans may have created an environment conducive to the emergence of the algorithm that provided a belief in the continuity of existence after death.

The consequence algorithm created fear, and fear created the sense of immortality and the idea of God. Or Gods, according to each religion. From this point of view, human intelligence, with its ability to calculate consequences, may have hindered, through fear, an activity of utmost importance for the species: hunting. Other animals don't have a consequence algorithm as complex as humans do; their fear is more in charge of their momentary instincts, without involving planning or premeditation. Humans who believed they would go to a better place after death were clearly braver than non believing humans. Humans who believed that they would be protected by

175

the Gods during the hunt were also braver than non believing humans. To believe in a God who protected them and in a life after death was surely a great convenience. The hunter with faith was confident, calm, and courageous. He fed his family well and helped his fellow tribesmen. A faithless hunter was a cowardly hunter. He did not provide his family with meat, the source of the proteins necessary to sustain the new, expansive brain that developed in these hominids. He had no friends and was frowned upon by the group. In an environment of hunter-gatherers with human intelligence, both the sense of immortality and the idea of God were widely advantageous properties.

All the religions I've heard about are based on immortality or life after death, as the reader may wish. It is clearly a fruitful behavior that originated more than 50 thousand years ago, because it occurs in every human society, from the Europeans, to the Africans, to the Asians, the Australian Aborigines, and the American Indians. Religions and the idea of God exist in all clusters of modern humans. As far as I know, there is no exception. All modern humans to date have some kind of a religion, usually with one or more Gods responsible for creating everything. Somehow, it is always thus.

Because it is a universal characteristic, I imagine that it was established shortly after the Toba eruption, in the early days of the emergence of human intelligence, in the extreme south of the African continent. Therefore, when modern humans began to leave Africa and spread throughout the world, they carried with them the sense of immortality and the idea of God.

From then on, the population of modern humans began to increase exponentially, with internal disputes increasing, in the beginning between tribes and then between cities. A new phenomenon appeared on the planet: great wars, which required courageous soldiers, making modern humans even more dependent on religions, by then controlled by kings and priests with their power politics. The idea of God came to be used as a form of power, even though it had arisen through evolutionary rules, to my understanding. It is the case of an instinct producing cultural results. Kings declared themselves representatives of God or the Gods. Clearer examples are the following: the Egyptian pharaohs and the Roman emperors. Some really thought they were gods. The Egyptian pharaohs were so preoccupied with the afterlife that they built immense, palatial tombs, the pyramids, where they imagined they would live forever. Caligula, the Roman emperor, thought that he was a God until they put a sword to him. Nero, another Roman emperor begged a slave to kill him when he discovered that he was not a God. The image of God became a copy of the court of kings, with princes, princesses, and queens. This strategy has been so successful that it has perpetuated itself through the ages.

Under the mantle of religions, these characteristics soon came to be used politically to dominate people. To this day, it seems to continue like this, only using mass media such as television and the Internet. Churches are founded every day in all parts of the world and use the name of God to ask for donations, not always spending the collected money for the good. I think that the initial cause of the emergence of the sense of immortality and the idea of God giving courage to humans and quieting their doubtful minds about the future was far more noble and honorable than that of propelling the creation of new religions, as is currently the case.

However, it is good to clarify some questions. My idea of the emergence of God does not in any way prevent one or more Gods from actually existing. It may seem that I am being politic, wanting to please both religious readers and atheists, but that is not so. My goal of sincerity is still valid. In this book, I do not treat religion as a belief, but as a successful human trait. I do not discuss the controversy of whether or not God exists, because this is a matter of faith, and here I'm dealing with science, which involves logic and evidence. I deal with the fact that most humans think that they are immortal and that there is a God or Gods, although today a large number of people have expressed doubts about it.

In any case, any theory about the origin of humanity can be seen as the way God created humans. The explanations of each religion stand as explanations presented according to the understanding ability of humans of the time. My theory does not escape this loop. God could have triggered the Toba volcano to erupt, which, according to my explanations, was the trigger for the emergence of human intelligence, which provoked fear and saw the advantage in the belief in immortality and the idea of God, to succor humans in their anguish, and, finally, to testify His existence. I cannot avoid explanations like this. In any case, my theory points to the emergence of humanity without the necessity of the existence of a God or of several Gods because it develops a reasoning that the meaning of immortality and the idea of God are faculties that have arisen with human intelligence and not by the interference of a deity. That is in fact correct: here I do not question religions. I present theories on the development of living beings, particularly modern humans.

18

The female orgasm

Another human characteristic that is complicated to treat, perhaps even more complicated than the sense of immortality and the idea of God, is the female orgasm. I reflected much before dealing with the subject, and even thought about avoiding it because I know it can offend susceptibilities. However, it is very important for understanding the procreation of modern humans and therefore for a better understanding of my ideas about the emergence of human intelligence.

I begin the discussion with males, more specifically, all male animals, at least the ones I've studied or with which I've had direct contact. After all, by drawing a parallel, it's easier to understand that similar patterns may well arise in the two sexes for entirely different motivations and in absolutely separate epochs. Adultery is an example of this. It occurs in both sexes, but it was established for distinct reasons and in distant times, as I will analyze later in Chapter 24.

Most males of the many species I have observed clearly demonstrate great pleasure during the sexual act and an incredible willingness to practice it, so much so that they're willing to kill, die, and rape for it. This uncontrollable desire and this great pleasure seem to guarantee the breeding of animals with reproductive systems similar to modern humans.

It's very difficult to know for sure what an animal is feeling, but from all that I have seen, including direct observations, films, documentaries, and

179

descriptions in books and other texts, I have never been able to detect outward signs of pleasure during the sexual act that resembles, at least minimally, the signals emitted by some human females. At most, what I have noticed is a kind of measured expression or resignation. Desmond Morris, by the way, comments: "The female does not appear to experience any kind of climax. If there is anything that could be called an **orgasm**, it is trivial compared with that of the female of our species." (MORRIS, Desmond. *The Naked Ape*, p. 68). The sexual act of horses, donkeys, chickens, dogs, lions, and cats seems to provide pleasure only to the male.

Generally, "heat" is referred to the predisposition of females to an irresistible urge to mate with the male in anticipation of immense pleasure. I have never noticed this. In some species, I even noticed a great interest before copulation, but during and after I have noticed nothing similar to what happens with the modern human females who experience orgasms. In females of other animals, there seems to be a predisposition for copulation and a predisposition to pleasure in males. For females, copulation would be the purpose, and for males, the goal would be orgasm, and sexual intercourse would only be a means to achieve it.

Modern human coitus lasts an average of 4 minutes (in the case of Americans); 1 minute in gorillas; 15 seconds in bonobos;7 seconds in chimpanzees; and 15 minutes in orangutans (DIAMOND, Jared. *The Third Chimpanzee*, p. 86). These differences, however, do not seem to influence the sexual appetites of males. The average coitus time suggests more a relationship with the environment in which they lived when this time was established in their adaptations, rather than with the intensity of pleasure. Gorillas, chimpanzees, and bonobos probably lived in dangerous environments, and so they spent as little time as possible in coitus, something that made them susceptible to predator attacks. Orangutans and humans, however, must have lived, at the time this quality was fixed, in quieter environments with fewer predators, since they could afford to spend much time on an activity that usually leaves the animal vulnerable. Thus, modern human, gorilla, bonobo, chimpanzee, and orangutan males can have similar pleasure during the sexual act, regardless of how long it lasts. As the intense sexual pleasure of males during intercourse occurs in so many species, it is quite clear that it is a fairly old occurrence, perhaps established well over 100 million years ago, since it is also observed in birds, descendants of dinosaurs, as science currently accepts. It also proves to be a very successful project because, despite small variations, the original format has been maintained.

With regard to females, however, doubts remain, and opinions differ. Considering everything I have read, including daily news, the female orgasm in other animals is admitted even if subliminally. Considering only scientific texts, however, the scale weighs the other way, with most opinions understanding that the female orgasm in humans is very different from what the female orgasm might be in other animals. However, even in scientific texts, I have seen some vague references to female orgasm in certain species of monkeys and bonobos, both males and females of which practice homosexual relationships. Gibraltar apes and South American nail monkeys engage in very intense and time-consuming sexual activity when the female is in heat; however, and I have observed this directly in Brazil, in my home state of Piauí, only the male appears to really have an orgasm. The participation of the female, although at times very active, in no way resembles that of orgasming human females. I watched some videos about the sexual activity of bonobos, and the images and descriptions of the females' attitudes indicate more of a social contact than a pleasurable experience related to procreation. From what I have observed, the way bonobos practice sex does not allow the faintest assumption that such gestures and sounds may indicate that the female is having a human-like orgasm. Except for humans, female primates do not express sexual orgasm (MORRIS, Desmond. *The Naked Ape*, pp. 60 and 61).

The most common of the animals I have analyzed is the menacing male trying to have sex, and the female merely complying. Often, she is resigned or even terrified. Females of some species even violently attack males in order to escape coitus. Female dogs, for example, when in heat, usually do everything to avoid sexual intercourse. They end up giving in, but they do not seem to take any delight in the act. So much so that, after sexual intercourse, resistance increases, perhaps because of discomfort without pleasure as a reward. Even those who show preference for a particular male at the time of the relationship show no sign of satisfaction, indicating that the choice has only a direct genetic motivation, without regard to a potential orgasm as an attraction. Here, the shocking example of lions is worth repeating. When the leader of a pride loses the rank by being defeated or killed, the winner kills the cubs of the lionesses, and then the pride's lionesses, in addition to accepting infanticide, still offer themselves sexually to the victor. The goal is simple and clear: the winner, in addition to not wanting to raise the progeny of others, want to generate his own soon in order to pass on his genes. The lionesses are not offering themselves to the new leaders for pleasure but due to the imposition of evolution, in order to replace the children they have just lost.

Also, female cats show a desire to engage in sexual intercourse with males when they are in heat, but during the relationship this does not seem to result in any pleasurable consequences. The fact that they are mewing and lowering themselves to the males may indicate that they will experience great pleasure, but what is seen in the sexual act of the cats is the great activity of the males and the complete indifference of the females, interspersed with a reversal of aggressions, as happens in the mating of other felines such as lions, jaguars, and tigers. Nature, save for some particular case I do not know about in mating, pushes the female to the male and the male to the female, but reserves the pleasure of sex for the male.

Thus, giving pleasure to the male, the problem of procreation was solved. The male has the desire and is determined to pursue the female until she accepts it. Nature seems to be correct in this aspect. All animals are there to prove it, procreating and multiplying, demonstrating that this project of evolution or adaptation, although it seems to be unjust, has been very successful. At least until something occurred with human females that caused some sexual pleasure to appear, similar to or even superior to that of males. In this chapter, my purpose is to try to explain what happened according to the principles of my Human Intelligence Emergence Theory.

Although research on human sexual activities using questionnaires is absolutely unreliable, both when it comes to men and women, I will cite some loose data without exactly a technical commitment, just so the reader can have a sense that it is widely accepted that the female orgasm occurs only in a part of the population. The reason for the unreliability is simple: when it comes to sex, men and women are almost always dishonest, for various reasons. These reasons, however paradoxical it may seem, are justified. Ever since boyhood, I have heard this saying as a joke, one of very bad taste, by the way, but it is quite apropos: when a man says he slept with three women, it is because he slept with one, and when a woman says she slept with one man, it was because she did with three. It is the man wanting to demonstrate that he is powerful, and the woman who is virtuous, because this gives them desirable qualities for mating. Research does not always consider the natural and innumerable conveniences of humans when it comes to sex. I come to think that research with real results is impossible to carry out, given the ability of the human mind to dissemble when it's important to hide facts and situations, as is certainly the case with information related to the sexual particularities of each person.

A survey published on the Internet (generalization is purposeful, as I said earlier) shows that 72% of women said they feigned an orgasm a few days before the interview, and 55% of men stated that they perceive when women are faking, which already clearly shows that the female orgasm, although obtained by only a percentage of women, is very important for both women and men. There is much research on the subject, especially in magazines aimed at the female audience, with similar results: some women say they only feel orgasm with lovers and pretend with their husbands; in general, a woman's first time is a great disappointment; men are very concerned about whether women have orgasms with them; and women think men are insensitive and think only of their own pleasure. As noted, the female orgasm is a major concern for both sexes. The information converges so that, unlike men, who usually always have an orgasm, only a certain percentage of women can achieve one. This appears in virtually all studies, surveys, interviews (with women and men) and testimonials (of women and men) on the subject. And it is this fact that really matters to my theory, as I shall explain shortly. The divergence is due to the percentage that goes from a minimum of 10% to a maximum of 90% of women who reach orgasm. After a couple of years of simple notes of Internet data, magazines, books, etc., I reached an average of 23% of women who reach orgasm, but I think that the percentage is much lower. It should be noted that surveys generally refer to women between 15 and 35 years of age.

Why do women pretend so much? It is a mystery that points to several solutions, some that do not contradict each other and even converge and complete each other. I do not wish to address them here because they involve controversial subjects that may offend susceptibilities and because they require much analysis that would take much focus away from the main objectives of this book. However, the facts show that orgasm is something that all women feel obligated to have, and that all men would like to know how to detect if women are feeling it with them.

Do men also pretend? The sexual problem of men is their erection. Erections cannot be faked. Either one has one or one does not. The male orgasm appears to be something so natural that its greatest dysfunction isn't when it does not occur but when it occurs quicker than it should. Premature ejaculation can be an embarrassment because it prevents the female from reaching orgasm.

It is not easy for women and men to solve these sex-related issues, precisely because they establish the rules of procreation. Although many psychologists and sexologists try to pass on the idea that all women can achieve orgasm, the facts do not show that this is true. And, even if there is the possibility of "teaching" women to reach orgasm, as many people claim, it must be something very difficult to "learn." At least a minimally efficient manual with widely accepted solutions has yet to be written. The proof of this is that this subject continues to yield and sell millions of magazines around the world, always with new ideas, indicating that there is still no solution.

Of the authors who mention that 90% of women reach orgasm, Morris stands out (MORRIS, Desmond. *The Naked Ape*, p. 66). But his data deserves a more accurate interpretation. He says that only 23% females reach orgasm by age 15 versus 80% males who reach it by age 14, although males begin the process of sexual maturation about one year later than females. However, he points out that in females this percentage rises to 52% at 20 years and to 90% at 35 years. This statistic seems very strange to me. I would therefore ask the pardon of Morris for the analysis that I will carry out below, which, if he has no scientific or statistical support, has at least one logical explanation.

It is precisely around the age of 14 and 15 that girls generally look beautiful, and their interest in the opposite sex arises. Also, at this age, hormones appear, and parents allow them to attend dances and even date. Unfortunately, without proper precautions, the first pregnancy often occurs around this age, especially in countries of the so-called third world. It really is strange that, with so many facts leading to sexual initiation in women, the orgasm, which is admitted to reach 90%, manifests itself so timidly, with an incidence of only 23%! It is stranger still that with little body change, from 15 to 20 years, it jumps by another 29%, rising from 23% to 52%. Even stranger still is the next rise of 38%, rising from 52% to 90%, when there is nothing to indicate in this sense, especially when it is known that even the pregnancy process becomes risky during that period. It should also be noted that, as Morris admits, these data are all derived from recent studies collected in North America (MORRIS, Desmond. *The Naked Ape*, p. 56), a wealthy and prosperous society that suffers great influence from extreme market capitalism where personal image is overvalued. It is not difficult to imagine the influence that this type of culture should have on research related to the female orgasm when it is clearly known that women feel deeply inferior when they are forced to confess their inability to feel the sexual pleasure that only a few achieve. I am neither blaming those responsible for the surveys nor questioning their data. In general, the results must be correct. The problem is knowing whether the interviewees are speaking the truth, when this truth is not confessed even to the pillow itself. When this truth puts them in a situation of absolute inferiority, which overshadows even the extraordinary value of beauty with an extremely cruel phrase: "She's beautiful, but cold." Thus, I propose that the age group especially influences the interviewees' answers. At age 15, still hopeful of one day feeling orgasm, she confesses her inability. At 20, even with a little perspective, only in some circumstances does she admit her lack. From then on, practically disillusioned and much more experienced, it is absolutely normal and even expected to swear that she feels orgasm. And she doesn't confess the truth to anyone under any pretext. She may even believe, or be deceived into thinking, that those small bouts she feels are an orgasm.

And don't let the reader think that I am discriminating (in the bad sense) against women, because men also lie a great deal when it comes to sex, as I have already mentioned. But the phantoms of human males are different: impotence and the inability to cause orgasms in women. It is incredible how most men begin to divulge an exaggerated sexual performance with their advancing age, when any layman knows that what happens is exactly the opposite. It is very common to hear young human males saying they have gone to bed with countless women, when often they have not even had sex. Demonstrating ability is the point of convergence of both sexes.

And what is in fact this female orgasm that only a few women can have and many pretend to have? I find it difficult to define because, as a man, I cannot experience it to evaluate it. And even if I could, I don't know if it would be very useful for an account, for I have never exactly understood the explanations of the women who seem to feel it. In any case, I'll write a description, based on books, interviews, articles, informal conversations, etc.

It begins with small moments of pleasure, which always increase in intensity, until reaching a climax where an apparent cerebral lack of control occurs that can provoke unexpected reactions of physical and verbal aggressions (sadism), and desperate requests of aggressions against herself (masochism). Then, there is a prolonged phase of relaxation, at which time the woman becomes more caring and helpful than she normally is. These are the most common manifestations; however, they sometimes do not occur and can be replaced by less or more pronounced and aggressive ones. It can happen on several occasions, depending on the person: with vaginal or anal penetration; with contact between vaginas; with solitary masturbation; with masturbation by the partner; with masturbation using an apparatus; in a dream; or by simply rubbing the vagina on cushions or pillows. The vocalization of both partners is also very important for the beginning of the climax of pleasure, if I can refer to it in this way. Even women who really appear to have an orgasm have a hard time describing it, as if it were something forbidden to say to their late-night companion or friend. It's an enigma!

At this point in the book, I think it necessary to offer a clarification. Or a reminder. I'm talking about only the science here, about logic and not ideology, misogyny, feminism, or racism. I do not defend minorities, nor attack them; so, no politics is involved at all. I can and even may be making mistakes, but I assume they are errors of logical interpretation. I may even be influenced by my philosophical thoughts, but I have tried hard to pursue neutrality in this regard. Thus, the female orgasm here is restricted to a single fact: the woman takes pleasure in the sexual act, here understood in its more general form, that involves several situations of intimacy already mentioned before.

Some authors relate the female orgasm with physical manifestations that normally accompany it, such as involuntary and rhythmic contractions of the striated circumvaginal pelvic muscles, the contraction of the uterus and anus, and fluid excretion through the vagina. As these manifestations occur with some monkey females during intercourse, these authors determine that these females have a sort of orgasm. I do not think so. Many human females exhibit these manifestations at some point in their sexual intercourse, and yet they cannot feel the female orgasm I am talking about, which refers exclusively to the intense pleasure that some women feel.

Having made these explanations, I believe I can state my suggestion regarding the emergence of the female orgasm in the way it manifests itself in modern humans today. The reader should note: by my reasoning, the female orgasm is not learned. It is not cultural. It is a characteristic. It may even be developed, but never learned. A woman is either born with the possibility of orgasm or she isn't. If there is such a possibility, she may one day be able to discover a way of feeling it. Otherwise, forget it. After all, not all the females of the other animal species have this peculiarity, and they continue to live and reproduce normally.

As I explained earlier, I suggest that in human evolution (from 2.5 million years to about 75,000 years ago), the brain first grew, causing an increase in the pelvic opening. Then a sudden decrease in the opening led to prematurity, which solved the problem of birthing a large brain, passing through a new pelvis, now a little narrower (around 75,000 to 50,000 years ago). Prematurity occurred precisely to allow babies to still be born with the small, incompletely developed brain inside the malleable skull, and thus be able to pass, even if it does so tightly, through the birth canal. This process, in my opinion, provoked the emergence of human intelligence. However, it has had a side effect to this day: in order to give birth to our babies, human females almost always go through a painful, torturous, and dangerous process that sometimes results in death. The fact is that certain evolutionary changes, especially those that are very urgent, end up being traumatic, and it may have begun to occur that some human females had a narrow pelvis, but without enough prematurity in the pregnancy so that the baby's head could pass through the birth canal, thus making birth impossible. I imagine that when hominids of the southern tip of Africa were becoming intelligent, between 75,000 and 50,000 years ago, there was a moment when the consequence algorithm arose in their brains, the same as I have already said that influenced the emergence of the idea of immortality and of God. For human children, birth came to be seen as very dangerous, and it became very natural that a female, possessing a consequence algorithm, to find that her mother, aunt, sister, neighbor, or friend had intercourse, became pregnant, and gave birth with great suffering or even death. What was to be expected in such a situation is that females would avoid sex, fearing the painful and dangerous consequences related to childbirth, and the human population would therefore begin to decline, which in theory would put the species at risk of extinction. I propose that one of the solutions engendered by nature was that, as an incentive to have sex, and therefore to procreate, it provided females with an extraordinary reward: female orgasm. Thus, the possibility of this extraordinary pleasure could overcome the fear of facing such a painful birth and of absolutely unpredictable results. So unpredictable that not even today, with the advances of science, it is possible to assume, even with minimal certainty, whether a birth will be normal or cesarean. This type of certainty is only found during labor, when, most of the time, the female human has endured cruel suffering. From the intensity of the female orgasm, I imagine that human female births, in the period when modern humans were emerging, must have been scary! In any case, I believe that the female orgasm originated from the emergence of human intelligence. To induce females to accept having sex with males at the risk of great suffering and even death, natural selection began to produce women with the capacity to feel an almost indescribable and unequaled pleasure.

Now, for the sake of justice, but also in the sense of fulfilling my purpose of sincerity, I'm going to comment parenthetically in order to admit that my arguments about the development of female orgasm end up justifying, if only partially, the fact that, according to sources cited by Morris, 29% of women say they only reached orgasm between 15 and 20 years of age, and 38% between 20 and 35 years, while only 23% say they had achieved orgasm before age 15. The reasoning is based on a widely accepted argument that considers that a lived experience has a much stronger influence than the observation of events happening to third parties. A female who had suffered the afflictions of birth, therefore, would have even more grounds for her refusal to accept sexual relationships than those who had observed the suffering of other females. Thus, the emergence of female orgasm at ages in which human females already had children would serve to make them risk once more the consequences of intercourse, facing the dangers and suffering of childbirth. This makes sense. But I still think that the data is inflated, as I explained above, when I discussed the value women give to their ability to have an orgasm. In any case, these are not conflicting arguments. On the contrary, they point in the same direction, leaving the divergence reduced to a quantitative value.

The explanation for the female orgasm occurring in only a percentage of women is that this ability was still being established when modern humans, due to the cessation of the universal climatic effects caused by the Toba eruption, began to spread throughout the world, which would have caused an evolutionary break, motivated by the suspension of the genetic bottleneck effect that occurred in the hominids that lived in the restricted environment of South Africa, and that extraordinarily accelerated evolutionary changes. With a brain capable of surviving in any climate of the planet without drastic adaptive changes, modern humans conquered all continents, practically with the qualities and defects arising from the special evolutionary situation experienced in the extreme south of the African continent. Even with only a percentage of women having orgasms.

However, some changes could have and should have occurred after human groups left Africa. But now, as normally happens in nature, in an environment without genetic bottleneck pressure, they occurred at a much slower rate, and there may have been, especially in human groups that had been separated from each other for a long time, a tendency to reconcile the opening of the pelvis to the size of the brain, pointing to a reduced mortality rate at birth. This also indicates that the encounters between long-separated groups of humans tend to be disastrous, since one group may have followed the direction of a larger increase in brain size and the other a larger decrease in the opening of the pelvis, which could lead to fatal incompatibilities already at the first crossing of a male human from a larger brain group with a female human from a narrower pelvic group. Of course, deliveries without problems are also expected when incompatibilities contrary to the example I gave occur, but even in the first generation of intersections between groups, depending on what characteristics the children have inherited, problems will begin. Maybe someday I will have access to data that will show if I am right about this reasoning.

19

Hidden ovulation and constant receptivity

In addition to the orgasm being restricted to a percentage of females, modern humans possess, in my understanding, other singularities brought about by the emergence of human intelligence and more directly related to procreation. Among these are occult ovulation and constant receptivity. I will discuss them together, because it seems to me to be a more didactic approach, and I also understand that they came about at the same time, influenced by the newly established division of labor. The hallmark of hunter-gatherer societies is that men were generally in charge of hunting, and the women were in charge of the gathering, entailing an ecological niche that combines the best of the two activities: meat protein and plant food reliability (RIDLEY, Matt. *Nature via Nurture*, p. 35).

Let the reader imagine a few hundred hominid groups with about 100 individuals each, living in both nomadic and fixed settlements in the extreme south of Africa, always in the vicinity of the ocean coast in order to facilitate sea-hunting, as hunter-gatherers approximately 65,000 years ago, when human intelligence was finally turning into something like the intelligence of modern humans, capable of sending an person to the moon and launching spacecraft out of the solar system. It is logical to think, based on my ideas about the emergence of human intelligence, that females, unable to hunt due to the constancy of complicated pregnancies caused by the narrowing of the pelvis, increasingly engaged in obligations attached to the place of residence, even if temporarily: how to take care of children and food, collect fruit, and even hunt smaller animals, probably with rudimentary traps. This type of social organization, clearly hunter-gatherer, favored monogamy (or, at least, moderate polygamy), because it provided children with the constant care of the mother and the necessary proteins from the father's hunts. While the rest of Africa was recuperating from the super volcano Tuba's eruption and human males, in groups, already ventured for many days to hunt larger animals in the interior of the continent (although sea lions still were the main source of food), evolution continued to push the narrowing of the pelvis and prematurity, creating an environment conducive to the emergence of human intelligence in an extraordinary bottleneck effect, which functioned as a change accelerator and a fixer of new characteristics.

In this situation, propagating ovulation with swelling, odor, or a coloration in the genital region, as with most animals (DIAMOND, Jared. *The Third Chimpanzee*, p. 87), would only create social conflicts because a female could enter heat when her companion, so to speak, was hunting, and the males who were close by, attracted by the smell, could come into conflict with each other and with the companion, who could return at any moment. The occultation of ovulation was, according to my reasoning, an imposition of the beginning of monogamy to guarantee the necessary stability for the survival of that strange new species that appeared in the extreme south of the African continent.

This happened in a similar way with constant receptivity. A type of system with marked monogamy, where the male would invariably be absent in order to hunt for and bring in such valuable proteins, would not be able to establish itself if the female had sex only when she was ovulating, because the probability of this occurring when the male arrived would be small, especially in certain seasons that required much time for hunting. Constant receptivity made the female available to the male almost any time he arrived, favoring monogamy, improving the chances of survival of the children, and even helping in the social balance of the group. Sex ceased to be restricted to procreation and also became a stabilizing function of the family. Perhaps the fact that the human male delivers the meat to the female human, and then has sex with her, has created a sort of cause and effect relationship that has led to female prostitution (DIAMOND, Jared. *The Third Chimpanzee*, p. 91), which has always been criminalized, but which today is even defended as a

191

woman's right. But I am not saying that female prostitution may have arisen unilaterally by obeying the evolutionary rules of Darwin's Natural Selection. Human males seem to enjoy extraordinary satisfaction from buying sex, and in general, they feel superior for it. In this way, should any moralist want to look for guilty parties, he should divide the guilt correctly between the two sexes. Female prostitution would have emerged through a two-way street and would even have noble motives for women: securing such valuable proteins for human babies.

20

Menopause

Menopause in human females, here understood as the loss of the ability to bear children at a certain age, is a singularity: our closest relative who has it is a tiny Australian marsupial, and it is the male that has the characteristic (DIAMOND, Jared. *The Third Chimpanzee*, p. 71). Thus, both we and the Australian marsupials experience menopause, but we are absolutely unique in our type of menopause, so to speak: in us, it occurs in females, and with the marsupials, in males.

As with every singularity, menopause also deserves an explanation. The most accepted, at least to my knowledge, is the Grandma Theory. According to it, human women lost the ability to procreate from a certain age in order to take care of their grandchildren. Their care ensures their genetic continuity more than having more children. These children would dispute meat proteins and other sustenance with the children of younger women, for example, of their own daughters, as I will explain in chapter 23, **Mating preferences.** In this chapter, I try to demonstrate the preference of human males for younger females. It makes sense, even though it requires an indirect consequence of evolutionary advantage, the effects of which do not seem to be universally accepted.

I don't have a concrete idea to explain exactly the loss of the ability to procreate in females, but I present a reasoning based on the way I imagine hominids developed in that extraordinary period when human intelligence and many other specificities arose, that may explain the phenomenon.

I believe that the vast majority of males lived at most until 36 years of age. And for this I take as a base the simple fact that, to survive, hominids hunted swam, especially after the eruption of Toba, and/or ran (endurance hunting), beginning from when the African interior was once again habitable. And modern humans are only able to perform these activities efficiently until approximately 36 years, as soccer championships around the world demonstrate. Players over the age of 33 already begin to emphasize technique over physical training. A breathless hominid without muscles for the hunt would be a nuisance and would end up being killed by some beast or abandoned by his own companions. Women, perhaps because they developed activities that demanded less of the muscles, probably lived a little longer.

By this reasoning, in those times, what was expected would be for men to stop procreating at age 36, and women at something around 45 years of age, simply because they would be dead after that. Therefore, a breeding mechanism that continues to work after these ages would be absolutely unnecessary. But the reproductive mechanism of human females generally terminates at age 45, while males can function until their 90s, or as long as their hearts and muscles are strong enough to withstand the effort of sexual intercourse. Thus, there would be no special reason for the existence of menopause: simply, life expectancy increased beyond the time of reproduction of the human females. In fact, menopause wouldn't even exist. The phenomenon would be the increase in lifespan, the origin of which I will explain in the next chapter.

If human males come to live, for example, to 150 years, they might be surprised that their reproductive mechanism only works to 100 years. They will have to live their last 50 years without reproduction and probably without sex, even if science can keep the heart and muscles functioning normally well.

Obviously, the ideal would be for all organs of the human body to last until the eve of death, but this doesn't happen, as is easy to see: most human beings die because of the failure of a single organ. And death almost always destroys countless organs in excellent condition, so much so that lately they are used in transplants with great results. So, I think it doesn't make much sense to say that menopause has evolved or developed over the past 40,000 years. Perhaps it is more creditable to say that the increase in the lifespan of human females to 80 years in the last 40,000 years has made them live 35 years without the ability to procreate, which they usually lose around 45 years of age.

21

The increase
in lifespan

As I said in the previous topic, the males of those hominids who lived in the extreme south of Africa, before the eruption of the super volcano Toba, had a lifespan of at most 36 years, when mainly their lungs and their muscles no longer had efficiency in the craft of hunting. A hominid male, without the strength to hunt or defend himself and his companions, was no good at all: he became a nuisance. And almost certainly he would have been discarded. Some must have been be killed by their fellow men, but there is fossil evidence that most hominids of those times were devoured by beasts.

With the emergence of human intelligence, modern humans came to have a brain with algorithms to solve many problems and with a great capacity to store data. If before they could recognize their fellow tribesmen, housing caves, hunting grounds, and the sounds of animals, they now recorded practically everything they saw and what their algorithm-filled brains had deduced. Thus, old men began to be useful, for they kept information of various situations that had occurred in years of experience when most of the young men of their tribe were not even born. If the region was going through a great drought, who knew where to find water in such a situation, since only he was alive in the last drought? The old man. If a strange snake bit a tribe member, who knew what to do? The old man, who knew exactly which herb to put on the wound, for he had seen the same situation several times. When a tribe was preparing to invade another village, who knew how to organize the defense? The old man, who, though not the chief, was, with his brain full of information, the chief's adviser and helped with tactics.

The experience of time came to have value and no one had more experience of time than the older members of society. Human intelligence turned useless clutter into indispensable information for the success of the group. Personified as counselors and healers, they came to be highly valued. Protected by the young and strong and cared for by the females, they became increasingly powerful, even leading their tribes. And the greater their ability to store information, retrieve it, and use it for their others in difficult times, the more important, necessary, and prestigious they were. I imagine that women followed a similar path and were living more and more, helping especially in the orientation of youths and in the raising of grandchildren. But I argue that males have led this race to the achievement of a longer lifespan, precisely because of the division of labor imposed by complicated and dangerous childbirth and the emergence of human intelligence made them responsible for the most complex part: hunting and the safety of the group, leaving for the females the gathering and domestic activities. Today, females have a life expectancy greater than that of males, but research shows that this is more because of better healthcare, especially because they seek more medical services, rather than because of any genetic motive.

22

Imperfect monogamy

Following the reasoning of the previous chapter, I propose that the difficulties, both in childbirth and in caring for the baby, led those hominids who survived in the extreme south of Africa to a clear division of functions between the sexes, with males hunting and females reaping, which induced males to mate with only one female and females to mate with only one male, leading to monogamy. This behavior occurs in most birds, but in few mammals (DIAMOND, Jared. *The Third Chimpanzee*, p. 99). In order to better illustrate the subject, I will now treat the procreation behavior of some species evolutionarily close to humans, beginning with primates:

• Common chimpanzees live in nomadic or semi-nomadic communities of up to 100 very promiscuous individuals, composed of several females and a smaller number of males, with copulation, depending on their social position in the group, invariably achieved through brute force or agreements with other males;

• Pygmy chimpanzees (bonobos) also live in nomadic or semi-nomadic communities and are quite promiscuous, such as those of common chimpanzees, but in a smaller number, up to 40 individuals, without aggression among males, and with a slight social leadership of the females, who are receptive to sex during most of the month, indicating the use of sex as a mechanism of socialization;

• Gorillas live in harems, led by a dominant male who expels its own male offspring as they grow, adding to the harem daughters who do not flee;

• Adult male orangutans are solitary, encountering females only for mating, in a behavior similar to that of cheetahs, about which I will speak later;

• Adult gibbons are isolated practitioners of territorial monogamy, with couples living only with their children, who are weaned at age three and leave their parents at the age of seven. There are no recorded cases of adultery;

• Gibraltar monkeys live in flocks, and females, when in heat, copulate with all the monkeys in the group, spending around 17 minutes each time;

• Lions live in nomadic prides, in harems, with one or a few males in command and with the right to mate with several females of the pride. When the cubs grow, the females are incorporated into the group and the males are expelled;

• Cheetahs are solitary, each male dominating a large territory, and only meeting the female to mate. The raising and caring of offspring are entirely the responsibility of the mother, except that the males seem to recognize their children, not attacking them when the mother enters their territory;

• Wolves live in groups, similar to lions, but led by a couple, whose components are denominated by the alpha male and alpha female. The alpha female enters estrus once a year and only procreates with the alpha male. A male's attack on one or the other is rejected by both. The whole pack helps in raising the young.

As the reader may have noticed, perfect monogamy is a rarity among mammals, practically restricted to gibbons, which have a lifestyle that is totally different from that of modern humans and live in solitary couples with only their young. We, who live in large groups, have a high rate of adultery, as I will explain later. The neighborhood, as it seems, doesn't help much in fidelity. We will only find situations similar to ours in seabird colonies such as those of seagulls and penguins, which live in large groups but also in pairs of male and female. I think the common point is childcare. Non-terrestrial birds generally dedicate themselves exclusively to the raising of their offspring, from the laying of eggs until they learn to fly with ease. The state of dependence of our children at birth is very similar to that of non-terrestrial birds: in both cases, for example, babies are born without knowing how to use the main means of locomotion of the species. The offspring of non-terrestrial birds don't know how to fly, and humans don't know how to walk.

As I mentioned at the beginning of the chapter, those humans were led by events to favor monogamy: an extreme need causes a narrowing of the pelvis, which hinders childbirth, leading to prematurity, which requires greater care of offspring and fosters the emergence of human intelligence, which established a semi-nomadic life, with a division of labor in which males were left with more external functions, such as hunting and protection of the territory, and females had the more domestic functions, such as fruit collection, caring for temporary or fixed housing, and child rearing.

All of these events, probably occurring at about the same time, including those related in List Four in Chapter 15, created a situation that clearly indicates a great advantage for those children whose parents tended towards monogamy because that behavior provided daily care and constant food on the part of the gatherer mother, besides the protection and proteins brought by the warrior hunter father. A lasting monogamous union in a situation such as this would only result in benefits for offspring, the prime objective of evolutionary forces. With so many advantages to monogamy, the emergence of male and female adultery in societies of hunter-gatherers does not seem logical. But, unfortunately, it does have logic. There is logic for male adultery and another for female adultery, as we shall see later.

23

Mating preferences

It is not easy to talk about human mating preferences, especially if one wishes to be honest. After all, this topic is about our mothers, our fathers, our sisters, and our brothers. Researching to learn what our father or mother found attractive in the other in order to mate is a sometimes a disappointing task. Unless, of course, we use our imaginations, as modern humans tend to do.

Could it be that our mothers were chosen by our fathers because they found them to be hardworking, upright, and honest - in short, special beings to help raise us? And were our fathers chosen for similar reasons? One does not need to live long to learn that these choices occur very differently with respect to the qualities valued by each sex. There are similarities, of course. It is in these similarities that we find more commendable faculties in accordance with our cultural concepts: both males and females want intelligent, cooperative, trustworthy, and faithful companions. In addition, they want to relate with beings similar to them (DIAMOND, Jared. *The Third Chimpanzee*, p. 117), in a kind of reverse xenophobia. The differences in desired attributes are where the problems with human cultures appear.

Research conducted in the United States and Europe around 1950 pointed to the coincidence of these preferences in the two sexes, but it also shows some differences, among them that women valued their partners' financial prospects, and that men favored women who were young and beautiful.

This result was disregarded by the psychology community, especially the part regarding the differences. The argument was that this reflected only the capitalist culture of the time in the United States and Europe, which supported the maintenance of the home by the man.

In 1986, psychologist David Buss, who had done research in Germany and Holland with the same results, then conducted a survey of 10,047 people in 37 different cultures spanning six continents and five islands, and the result was the same: men like beautiful, young, loyal women, and women like rich, ambitious, older men (RIDLEY, Matt. *Nature via Nurture*, p. 75). It was a very important study for science, and particularly for my theory, but it confirms only what any attentive person sees in any grouping of humans. Observing with an unbiased view, it is not difficult to see that, in general, men choose beautiful, young, and loyal women as partners. The aspect of youth in relation to the taste of males is very important. To cite beauty as a preference for mating even seems redundant, since it is linked to vivacity, as we see in all the descriptions of beautiful women from novels to advertisements: smooth, angelic skin; firm breasts; a childish and dreamy look; firm, round buttocks; a thin waist; soft hands; and so on. In prehistory, I presume with a great degree of certainty, that almost no woman possessed these characteristics after the age of 30. Today, with health care and fitness trends, it is evident that many women maintain their beauty for much longer simply because they look younger. Likewise, with regard to females' mating preferences, the ambitious aspect is linked to wealth, which in turn refers to the older and more powerful, for it is difficult to achieve wealth or power by being generous, selfless, and merciful while young.

Since these are properties of all peoples known today, according to the Toba strand of my theory, they must have arisen in conjunction with the emergence of human intelligence after the Sumatran volcano erupted 75,000 years ago. This was when humans were living as hunter-gatherers in the extreme south of Africa and were suffering from severe deprivation. They were forced to seek food from the rivers and oceans and were dependent on sea hunting before returning to a predominantly terrestrial hunting life, when the catastrophic results of the eruption and the humans of the time spread throughout the planet. These preferences would have been established in this period, and so, I need to explain them.

Those hominids who escaped the Toba catastrophe in South Africa were trying to survive in an exceptionally hostile environment, using ground-based endurance hunting and sea hunting through swimming in rivers and oceans. With the onset of prematurity, the emergence of human intelligence, and the prolongation of childhood, it became a necessity for the female to stay in caves or camps to care for the children and ensure daily food gathering. Already, the powerful brain that was emerging demanded extra proteins that vegetables did not provide. In this way, human males began to hunt more consistently to provide protein-rich foods essential for their offspring's brain development. For a human female, having a relationship with an experienced male hunter, who was usually more powerful, older, and bolder, would have been very useful because it would have kept her children strong and well fed. The male's age, therefore, would be a prerequisite for success in hunting, since the extraordinary brain that was becoming established required learning time to acquire the knowledge needed to hunt in such an unfavorable environment as that caused by the Toba catastrophe. It is not difficult to understand that females who chose older males had a good chance of raising strong and healthy children, provided by the proteins brought in by their successful companion. And here, perhaps, is the explanation for the preference for rich men found in research: successful hunters of those times correspond with the rich men of today.

The explanation of males' predilection for younger females is more direct and is, in my view, more related to the possibility of the female becoming pregnant, restricted both by menopause and, especially in those years, by complications from childbirth than by the ability to raise children longer. After all, activities associated with the collection and maintenance of caves and encampments were not essentially restricted by the aging of females, while menopause limited maternity.

In this way, those hominids who might have liked older women had no descendants; only those who liked younger women would have had offspring. So for human males, having a relationship with a beautiful (beauty being related to youth), young, and loyal (to assure their paternity) human female would have been of great use to have many children with a considerable assurance that they would be theirs.

Men and women are actually very different in their behaviors, even in their fears. Human males are haunted by the ghosts of (financial) bankruptcy and sexual impotence, inevitably accompanied by failure as a child provider and by the disinterest of females. Human females, however, are haunted by the phantoms of old age and sexual indifference inevitably accompanied by the loss of beauty and the disinterest of males. In evolutionary terms, a man who does not procreate and/or is unable to raise his children is a failure and therefore is treated coldly by females. And a woman devoid of orgasm and/or the beauty associated with youth is also considered a failure and treated with indifference by men. This becomes clearer when you imagine a hunter-gatherer society. More evident still is the world today, with females increasingly displaying their physical qualities related to youth in a clear demonstration of their sexual desire for males, and more and more males bearing outward signs of wealth, also in a clear demonstration of dependence on being financially desired by females. And both show or, at the very least try to demonstrate, that they are good in bed. The Internet highlights this every day.

In the film Troy, quoted in the Epilogue of this book, Ulysses, king of Ithaca, trying to explain the death of the young Greeks in the first battle on the shores of Troy, when the elders celebrated and divided the riches they had conquered, speaks to Achilles, the king of the Mirmitones and the greatest warrior of Greece, who always threatened to break the alliance with Agamemnon, king of Missenas and commander of the Greek army, who awaited him for a difficult conversation about who had really decided the victory for the Greeks: "In war it's like this: young people dying and old men talking." Also, to simplify the mating preferences of humans today, one can say: in today's world it's like this: women taking off their clothes and men buying cars.

There's nothing pejorative about this, neither in relation to men nor in relation to women. With humility and understanding, it is not difficult to realize that it is only nature acting through instincts. A woman, when taking off her clothes, demonstrates her beauty in the most complete way possible. Similarly, a man evidences his wealth through the car that he is able to buy. Thus, they put themselves on display, with the aim of attracting the opposite sex: women showing beauty, and men exhibiting power, because, in this way, they will be able to choose the best partners to have strong, well cared for, and nourished children.

24

Adultery
and private sex

Adultery, both male and female, is generally considered to be a moral defect. In a way, this is true at least when we compare it with the cultures established in most human groupings. It could be said that the dominion of the human male over the human female caused the emergence of cultures that do not accept or even criminalize adultery because it is always more related to the sexual and/or affective infidelity of females. The first legislation that criminalized male adultery was passed in 1810 in France and merely forbade married men from keeping a concubine in the conjugal home against the will of their wives. Meanwhile, various cultures had already criminalized female adultery for more than 4,000 years, such as the ancestors of the Jewish and Arab peoples, who condemned adulterous women to death by stoning or whipping.

However, women, in more modern times, so to speak, were also revolted by male adultery and somehow influenced the laws and customs, adjusting them to their own interests, desires, and satisfactions, which had little to do with the interests, desires, and satisfactions of the women of today. They struggled generally for the success of their children: Hatshepsut—sister of the Pharaoh of the time of the Exodus—for Moses (Egyptians/Hebrews); Agrippina for Nero (Romans); Olympia for Alexander (Macedonians); and Bathsheba for Solomon (Hebrews); among others. Feminism is a recent movement, and it is only now that it has been changing laws and customs in the sense of providing satisfaction to women regardless of the success of male children. Among some people, such as the Muslims, it has faced great resistance. Even the women themselves, who want rights such as the right to vote, vehemently reject the prohibition of polygamy and denounce exemption from the requirement to cover the face with a veil.

In this chapter, I avoid commenting on the moral or ethical aspect of the behavior. When that happens, I provide the stipulation that I think is fit for the understanding. I try to identify the instincts that wield influence in humans today to explain the causes of their origins. I know that some subjects may shock more sensitive people, who are accustomed to seeing everything through the lens of convenience but can find no other way to approach these important behaviors, particularly because of their connections with the emergence of human intelligence.

There is no need to prove that adultery exists. Every adult has heard so much about it that no evidence is necessary. Few motives have equaled it in terms of causing murder and human suffering (DIAMOND, Jared. *The Third Chimpanzee*, p. 99). Only wars and hunger appear to rival it. Music, movies, plays, books, the press, and gossip magazines devote so much space and time to adultery that strictly speaking, it would not even be necessary to mention some research or experience to admit that it is a human characteristic and not merely a transgression of the normal pattern, which would be perfect monogamy. Research on the subject is unreliable because almost everyone lies when they talk about adultery (DIAMOND, Jared. *The Third Chimpanzee*, p. 97) just as they lie when they talk about sex. However skillful interviewers might be, they are deceived by those who already have a lie prepared and who are trained for any occasion. However, I cite some experiments that Jared Diamond selected in his book *The Third Chimpanzee*.

Around 1940, an American physician (the author calls him Dr. X to preserve his identity) studied the genetics of human blood groups and collected blood samples from several couples and their children in a reputed hospital. The result was a surprise: almost 10% of the babies were the children of adulterers. A while later, many similar studies were done, both in the United States and in England, finding that between 5% and 30% of children were from extramarital affairs. Considering that several substances related to blood groups were not yet known, the real incidence of extramarital sex must have been much higher (DIAMOND, Jared. *The Third Chimpanzee*, pp. 97 and 98). If we also note that blood groups do not detect all of the children born of adultery and that most cases of adultery do not produce children, the result will easily rise to 50%. Finally, if we still evaluate male adultery to always be seen as greater in incidence than female adultery, the percentage of extramarital sex will exceed 70%. And here, we are talking about American and English discretions in the 1940s and 1950s! With such a high incidence, its significance is very great in human evolution, and its influence is extraordinary in diverse human behaviors. However, its study is painful and exhausting because after all, we are talking about the instincts and behaviors of our fathers, our mothers, our brothers, our relatives, and our friends as I mentioned in the previous chapter regarding mating preferences.

I propose that adultery, both male and female, arose while monogamy was emerging. It is not the case that monogamy came first and adultery later. According to my ideas, we have never lived in a period of perfect monogamy, such as, for example, the one in which gibbons live today. Our monogamy established itself as a package deal, together with its imperfections. The adultery of each sex, according to my reasoning, arose for a different reason. Again, when I speak of the cause of an attribute, I always consider that there may have been more than one although I use the verb in the singular to avoid expressions such as "among the main reasons" and "the most important." Therefore, other factors may have influenced adultery. Readers themselves may discover better solutions than mine or one that is the most important.

It seems simple and logical that human males should look for several females because only then will they have more children. A male who stays faithful to his female will have the possibility of having only one child per year. In 7 years of cohabitation, he will have a maximum of seven children. Sixty thousand years ago, it would probably only have been two as breastfeeding restricted the ability to conceive. However, a male who could mate with five females over 7 years could produce 35 children, an extraordinary evolutionary advantage. In this way, it explains why males "ruined" our monogamy.

With females, the story changes. It is useless for a female to have relationships with five males as she will continue to have a maximum of 1 child per year. The reader should be a little more attentive because female adultery, to my understanding, arose because of two advantages that occur simultaneously.

Here it is important to remember that the human brain weighs less than 2% of our body weight and consumes an extraordinary 20% of the body's metabolic resources (KLEIN, Richard G., BLAKE, Edgar. *The Dawn of Human Culture*, p. 121). A child, with a costly organ like this developing to house human intelligence, certainly had a great and constant need for proteins to ensure its growth. From this angle, one realizes that it would be an excellent advantage for the children that the mother has relationships with more than one male because it would secure more protein-providing hunters to contribute to their greedy brains. Simple mathematical logic can explain the origin of female adultery: two hunters bring more game than one, while three hunters bring even more.

This would be the first advantage: to have more proteins so that her children grow strong and healthy. The second would be to have one or more reserve males to supply meat if her mate died while hunting or in a territorial defense fight, which was quite common. In that case, female adultery would serve as a kind of "work accident insurance." Human females would clearly be using a quality-based strategy, betting on a few strong, well-nourished children against many weak and malnourished children, who would be a risk for the male strategy of having relationships with multiple women to have many children. With regard to adultery, human females seem to have used a more efficient strategy than that used by males as, at least in theory, the choice of future group leaders would remain in their hands. Healthy breeding was an important factor in a dispute where the decisive qualities were muscle strength and intelligence, preparing leaders who are highly dependent on a protein-rich diet in childhood. However, this had drastic consequences for the future of human females, transforming them into slaves of beauty and vivacity.

All animals that live in groups have public sex, whether monogamous, polygamous, or promiscuous, including the closest ones to humans such as chimpanzees, gorillas, and bonobos and even birds such as seagulls, which copulate in the middle of the colony. However, humans in general do not like to have sex in front of other humans. I think this particularity arose as a logical consequence of the imperfect monogamous system that we acquired after the Toba explosion. It is not difficult to understand that a male becomes violent when he discovers that his female is having sex with another male because this creates doubt regarding the paternity of the female's children, eliminating from practically one moment to the next all that he had invested for years to nourish them. In fact, this occurs daily around the world, resulting in many women being beaten or killed. While being reprehensible in all moral and ethical aspects, it is the logical response of the evolutionary process to the situation. A male who accepted this as normal would be a paradox because he would end up raising other men's children, and soon in competitive struggle, his passive genetics would become extinct, leaving behind the genetics of those who resolved the situation violently. This would result in the elimination of both the unreliable female and the probable children of the other male, either by direct action or due to the lack of care from the dead mother and from the father who would have abandoned them. As scary as it sounds, we are descendants of hominids who were murderers of adulterous females, not of hominids who were perspicacious toward female infidelity.

We can also presume the unhappiness of a human female on witnessing her partner in sexual intercourse with another female for this would jeopardize the supply of calories to her children as the hunt would also be divided with the other female and the other likely children. However, the difference is evident: the fact that the partner has a relationship with another female does not create doubt about her motherhood; her children are still her children. The logical reaction of the female is to find another male to mitigate the injury that the other male and his children would cause to her and her children. This has support today for it is much more difficult to see violent assaults committed by women who discover the adultery of their companions.

In any case, practicing adultery in public risked upsetting the balance necessary for the development of children. Male adultery, as it could lead to female adultery and its violent results, caused dire consequences. I propose, then, that private sex arose out of the need for adulterous hominids to have hidden sex.

In short: caring for children has led to monogamy; the advantage of having more children led to male adultery; the advantage of having more well-fed children gave rise to female adultery; and the need to practice hidden adultery gave rise to private sex.

25

Homosexuality

To me, male homosexuality was one of the most difficult human behaviors to explain. It would have to overcome a simple but seemingly insurmountable logic: most homosexuals do not reproduce, and physical or behavioral characteristics that hinder reproduction usually do not last. Since it was not exactly one of the aims of this book to justify all human behavior through evolution, I became accustomed to the idea of touching on topics without presenting a minimally viable solution. However, finally, I did come to a conclusion.

First, I try to clarify how I understood the discussion between opponents of the term "homosexualism" and those who advocate the use of the term "homosexuality." Some years ago, "homosexualism" was used more frequently, precisely because the gay community (which was also an accepted name—today, the most defended terminology is "homoaffective") argued that having relationships with others of the same sex was a sexual option, directly conflicting with a majority of the Christian churches, which considered it to be a curable disease. Now, however, homoaffective communities have come to criticize the use of the word "homosexuality" precisely because they no longer accept that the behavior is an option but instead affirm it to be a genetic imposition. They then go on to advocate the use of "homosexuality" to be more in agreement with genetic causes. That is, homosexuals, who formerly called themselves homosexuals because of a personal choice, started to call themselves homosexuals because of their genetics. Here, the two terms may be used because I think, in support of some scientific opinions, that the behaviors can be influenced both by genetics and how one is raised. However, I prefer to use the term "homosexuality," which refers to the characteristics of homosexuals, rather than the term "homosexualism," which refers to social movements that support homosexuals and their particularities. The relationship seems to be very similar to that between "femininity" and "feminism"; the first term refers to the peculiarities of human females, and the second one refers to the social movements in support of these peculiarities. Similarly, although many disagree because it is considered to be "out," this reasoning can be applied to the relationship between "masculinity" and "machismo" as well.

While studying other subjects, I found logic in some propositions that explain homosexuality—the behavior may have appeared as a birth control tool. In this case, a large population increase could have occurred in the extreme south of Africa, where our ancestors were besieged in the aftermath of the Toba explosion sometime after 75,000 years ago and before 60,000 years ago. Exaggerated population increases, however, have occurred in several species of animals without the emergence of homosexuality such as that observed in modern humans. The most cited case of homosexuality outside of humanity is that among bonobo females, who embrace each other, rubbing their genitals and squealing. Honestly, I have seen these scenes in documentaries, and they look more like an activity linked to social relationships, such as the habit of chimpanzees of removing the parasites of another, than anything resembling a sexual relationship. One day, I would like to directly investigate the sexual behavior of these apes to, finally, provide a complete and more certain opinion. Even accepting that homosexuality exists among the bonobos, it would be a singularity shared with humans, such as the menopause, which, as I have already mentioned, only appears in one living species other than humans: a tiny Australian marsupial. I have, therefore, come to think that male homosexuality in modern humans is also a singularity, and like most of our singularities, it may have some relationship with the emergence of human intelligence. My explanation goes in that direction.

Once again, I ask readers to imagine hominids living around 65000 years ago, practically trapped in the extreme south of the African continent, with a brain very similar to ours—full of algorithms to solve several problems, with large spaces for saving data, and equipped with an extraordinary computational system to process them. The new development is a great improvement in the general conditions of the continent, enabling increasingly distant and lingering incursions into its interior and the hunting of larger animals, which were certainly more difficult to slaughter. It is possible to imagine that hunting took increasingly more time and required the long separation of human males from their females because of the division of work in those hunter–gatherer societies. Constant sexual receptivity, which I referred to in Chapter 19, certainly influenced the sexual desire of males, which also became constant. It is logical to suppose that this model hampered hunts, with males frequently returning, eager to have sex with the available females. I imagine that under these circumstances, a trait such as homosexuality could arise because it would provide a balance, which would keep the hunters going about their activities. Having males who enjoyed sexually interacting with other males became an evolutionary advantage as this would make possible longer, more effective hunts, leading to the obtainment of more protein, needed to support the development of the brains of children.

What about female homosexuality? I think that the absence of males due to these long hunts, the beginning of the appearance of attempts at monogamy, and even the beginning of the female orgasm provoked proximity and cooperation among females that evolved into feminine homosexuality in the form that we know today, which, it should be remembered, is very different than male homosexuality. I will not dwell on these differences because there would be a need for more consistent information to express an in-depth opinion, which is not the purpose of this book. However, I cite only one so that readers understand the difference between the two behaviors. Most gay men go out in the evening looking for partners and pay for the sexual act, often with a stranger, which is hardly true for female homosexuals, who prefer more affective, long-lasting, and emotional relationships usually found in friendships. In any case, I try to show why homosexuality, as with many other natural human characteristics, can be influenced both by genetics and by how one is raised as modern science, or at least a part of science, now argues.

The genetic influence on animal behavior is fairly easy to observe. Dogs are a great example: cross ferocious dogs, and one will get ferocious dogs; cross docile dogs, and one will get docile dogs. It is obvious that the way they are raised will also have an influence, but the natural genetic influence is clearly noticed. Often, even careful raising cannot change those qualities.

This also seems to happen in humans. This influence has been well documented, including by various experiments using identical twins raised separately. Even for twins who have never met one another, the results show a great similarity in behavior, signaling a correlation of 62% on average, which indicates a very strong genetic influence but never determinism. It should be noted that this correlation drops to 2% for fraternal twins (RIDLEY, Matt. *Nature via Nurture*, p. 105). Several similar experiments have been performed, and the results are always very similar. Therefore, homosexual activists who defend genetic influence are not without a little reason.

At birth, the offspring of geese, ducks, and chickens will be fixed upon the first thing that moves near them and will follow it as if it were their mother (RIDLEY, Matt. *Nature via Nurture*, pp. 194, 195 and 196). I had the opportunity to observe this happen several times in my youth when raising chickens. Nearly all the breeders I met had a chick, usually rejected by the mother, that would follow them (the breeders) through the chicken coop. I have had several myself. Every time I walked into the hen house, the chick would beg for food and curl up at my feet. Austrian zoologist, ethnologist, and ornithologist Konrad Lorenz (Nobel Physiology/Medicine 1973 for his studies on animal behavior [ethology]), in 1935, scientifically described this phenomenon, which was called imprinting, including delimiting the timespan in which it can occur for goslings—between 15 hours and 3 days after birth.

I think of imprinting as a kind of window that opens so that a certain behavior is recorded according to the external environment. This way, at 15 hours of age, the "window" is opened so that the gosling's brain can record the physiognomy and other peculiarities of its mother, aiming to recognize her from then on in any situation in which it needs her. If, instead of the mother goose, there is a cat, unfortunately the result will be tragic. This seems to occur in the behavior of nearly every animal, including that of modern humans. In this case, genetics determines that during this period, the gosling will recognize its mother; however, this behavior will depend on which animal is in front of it at that moment. In my case and in that of other chicken breeders, as we fed the rejected chicks, they ended up identifying us as their mother.

I came to consider imprinting as an explanation for female orgasm; however, I did not find a strong causal link between events, particularly because of the lack of reliable research data, and I imagined that if there is a window to start the process, it must be caused by something very subtle, even if doubtful, as the research, because of the lack of sincerity in the answers, as I explained in Chapter 18, indicates that the few women who do experience orgasm feel it at different ages, between 13 and 30 years, without a common fact that could explain why this happens. One day, who knows, I might still write something about it.

Another interesting experiment happened by chance during World War II, when Germany invaded and dominated the Netherlands and there was a disagreement between the invaders and the invaded, provoking retaliation by the former against the latter, which caused a terrible famine in the country, leading to more than 10,000 deaths from starvation and affecting some 40,000 fetuses in gestation (RIDLEY, Matt. *Nature via Nurture*, pp. 198 and 199). In the 1960s, a team at Columbia University studied the data collected during this period, and among other expected results, they found a strange fact: infants who were in the last trimester of gestation when the food deficiency occurred became diabetics in adulthood even without cases of the disease in their families. The explanation is as follows: a fetus that suffers because of the needs of a poorly fed mother during gestation is somehow prepared to live in an environment of food deprivation, and the entire metabolism is equipped to harness and accumulate fat, which, as everyone knows, causes the onset of diabetes. This shows that even an external, pre-partum environment can influence human characteristics, with results detectable only 30 years later.

Thus, I think that homosexuality, just as in the case of many other physical and behavioral human properties, is influenced by both upbringing and genetics. One should not, therefore, think of a homosexuality gene. One should think of several genes that influence the individual toward homosexuality. At least in theory, those born with this group of genes that influence homosexuality would be homosexual regardless of how they were raised. In thesis, those who were born with all genes pointing to heterosexuality would be heterosexual regardless of how they were raised. Now, those who have a certain number of genes on each side would depend on upbringing for the definition of their sexual behavior. If, however, the genetic force were to fall more to one side, it would depend on a more contrary upbringing environment to modify this situation. It would be like a genetic tendency that could be influenced by upbringing, with two extreme cases, at least in supposition, of all the involved genes leading in the same direction toward genetic determinism, at least theoretically. I imagine there must exist multiple windows at various times when a child or adolescent may be influenced toward either heterosexuality or homosexuality. This is a position similar to that of Matt Ridley, who believes that human behavior should be explained not only by nature but also by nurture, that is, by upbringing.

Here are two important questions: what is the percentage of men and women who are homosexual, and when did this specificity arise? I have not found scientific books dealing directly with the subject, and I have had to settle for research results and the opinions of people based on their life experiences. As in all matters related to sex, research is hampered by the lack of sincerity of the interviewees. Because, in general, cultures do not accept homosexual behavior, classifying it as shameful and sinful, it is assumed that many homosexuals mislead interviewers and call themselves heterosexual. This presumption seems to be true because many of those who declare themselves to be homosexuals in surveys say that they continue to hide the fact from family members and even from friends. Thus, the percentages presented, 18% for men and 7% for women, based on the average of several surveys that I have noted, should be much higher—perhaps approximately 25% for men and 10% for women. It is true that these results are approximations, but I chose to quote them so that readers can get an idea. On comparing these statistics with the approximate result of 15% for the occurrence of female orgasm, readers might get the idea that may even be right, but for the wrong reasons, that male homosexuality is older than female orgasm because its incidence is higher.

However, readers should pay attention to the peculiarities of each attribute. Female orgasm seems to be a specificity that was intended for all women, and its occurrence can really point to the time that it emerged and spread throughout the world. Homosexuality, however, seems to have a limit of occurrence that does not detract from the necessary birth rate of humans, and its emergence remains truly indefinite. Thus, it is acceptable to say that the idea of God arose before female orgasm because the former is a faculty that reaches 100% of humans, while the latter is restricted to a small percentage.

However, a similar reasoning cannot be used with regard to homosexuality, which may have arisen before and has remained at the percentages of today by the delimiter factor.

Yet, it seems certain that it emerged in the extreme south of Africa at least 60,000 years ago because homosexual behavior was found in the peoples of the Great Balin River Valley, who still lived in the Stone Age, in the supposedly uninhabited interior of Western Papua New Guinea, discovered by the Third Archbold Expedition, which was led by Richard Archbold, on August 4, 1938. "Baruas" men exercised institutionalized bisexuality, even having a family, home, wife, and children and living with children in a large homosexual cohabitation. The Tudawhe had two-story houses, where women, children, single girls, and pigs lived below, and upstairs, which could be accessed by a ladder, lived single men and young men (DIAMOND, Jared. *The Third Chimpanzee*, p. 255). As these people arrived in Oceania 50,000 years ago from Africa, they certainly brought with them this type of behavior. There may have even been a change in the percentages of occurrence because of the need for birth control, which was caused by the great difficulty of finding food in this part of the island. Unfortunately, I was unable to obtain data regarding the female sexuality of the people of Papua New Guinea to know whether female orgasm was already present in the peoples who came to Oceania.

26

African genetic diversity

Now seems to be the appropriate time for me to digress and deal with a phenomenon, which, I believe, originated precisely at the time when human intelligence had already emerged and the first migrations began to take place—the great African genetic diversity.

Human genetic diversity is much more pronounced in Africa than it is in the rest of the world. This implies that humans in the rest of the world are genetically more similar to each other and to some Africans than they are to other Africans. That is, these other Africans are the humans with the most diverse genetics on the planet—and probably the oldest. At least this is what I understand from the texts that I have read about the mutations that occurred in both mitochondria and Y chromosomes. This requires an explanation, particularly when I argue that humanity came into being in one place, southern Africa. If mankind emerged in a place as restricted as extreme South Africa in a time as short as 15,000 years and spread throughout the world, the expected result, in my opinion, is that there would be a similar genetic diversity in all of the planet. However, this does not happen. Many studies are confirming great genetic diversity in the populations of the African continent, particularly in the south.

According to my thinking, when the catastrophe of Toba devastated the planet, or at least part of it, only hominids in Northern Europe, who were already Neanderthals, or gave rise to them, survived. In the extreme south of

Africa, probably hominids very similar to Idaltu gave rise to modern humans. I even admit that more hominids may have survived but were extinguished due to some ecological cause or were decimated by the descendants of one of the two groups.

From then on, trapped both by the hell that the rest of Africa had become and by the oceans in the extreme south of the continent, facing an extremely hostile environment, the hominids that gave rise to modern humans suffered a huge evolutionary pressure that caused a rapid increase in genetic diversity. However, because of the difficult and restricted environment in which they lived, the new adaptations were so necessary that those who did not possess them became extinct in a few generations, thus liquidating the genetic diversity that had just been created. This way, hominids were maintaining, and perhaps, even diminishing, genetic diversity. I imagine that life would have been very difficult for the handful of survivors who were very similar to us. I also wonder how many times they came close to extinction.

The fact that all survivors were from the same location further increased the effects of the genetic bottleneck as minor changes were absolutely necessary for that limited environment. It is no exaggeration to say, by way of illustration, of course, that in a small area in the extreme south of Africa, a "laboratory" had been established to produce intelligent humans from a group of hominids existing more than 75,000 years ago.

When in Chapter 8 I explained my theory, pointing to the eruption of Toba as the event that brought about the emergence of human intelligence, I defined it for didactic purposes; however, based on the various dates of other related events, particularly on the dating of the oldest modern human fossil found in Oceania, the emergence of humanity occurred around 60000 years ago. Therefore, it is logical to also think that around that time, the rest of the African continent improved in its inhabitability and that some of those modern humans began to leave the southern tip of Africa to populate the African continent first and then to settle the rest of the planet. Thus, two events must have occurred during the same space of time: migrations across the continent and decreased evolutionary pressure on the pelvis. They were caused by the same reason: the improvement in the living conditions of modern humans in southern Africa and in the rest of the continent.

Consequently, when the first migrations began, there was still sufficient evolutionary pressure for more pronounced changes, which, in a few generations, led to a great genetic divergence between those who came first, **the probable ancestors of African peoples of marked genetic diversity**, and those who were then likely to be **ancestors of the rest of the African population and of the peoples of the world**. The time gap between the two or more migrations would explain the genetic diversity found today. Each group of modern humans that ventured inland carried a different genetic charge, sharply modified by an environment that still had genetic bottleneck patterns. These first migrants settled in isolated and more livable regions, maintaining for thousands of years the genetic load brought on by the small evolutionary pressure that the mild climate of the planet now provided. Subsequently, it is possible that they procreated among themselves and originated the peoples who have inhabited Africa for tens of thousands of years. Genetic studies, however, show that the San people, who inhabit the Kalahari Desert in the south of the continent, have the greatest genetic diversity among humans, indicating that they have remained isolated from other peoples since the migration to the southern tip of the continent. This suggests that migrations outside the continent must have occurred last, when the bottleneck effect practically no longer existed as I show in the map in Figure 15.

This is a suggestion, not a guarantee. If, for example, a tribe is found in Papua New Guinea with greater diversity than that of the San people, the map in Figure 15 should be changed, revising the first migration as the one that came to Asia, which would explain the hypothetical tribe we imagine to illustrate the reasoning.

However, as current data indicate that the greatest diversity occurs in South Africa, which is very close to where I suggest human intelligence and hence modern humans emerged, there is a possibility, however remote, that there is still some human group, perhaps some isolated tribe, with a genetic charge totally coming from some humanity before the humanity that was distributed throughout the planet if there are still isolated tribes in Africa as there are in Papua New Guinea and in Brazil. Readers should note that I am not saying that they are humans who originated before modern humans but only that they are modern humans who first migrated and for some reason, were isolated. This is not to associate them with humans who still had not

START OF MODERN HUMAN MIGRATIONS

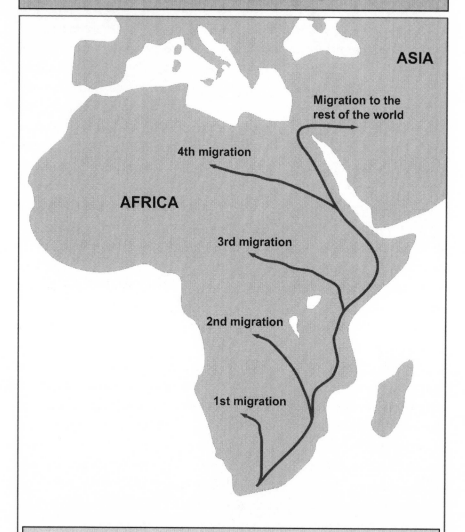

My placement of four migrations to inland Africa is not a suggestion but rather a simple didactic device seeking to show that as genetic studies have indicated, Africa was probably populated through more than one migration.

Figure 15

fully developed intelligence because nothing indicates that the humans who migrated later had similar, less, or more intelligence than the first migrants.

After this comment, it is possible to compose a picture of the distribution of modern humans all over the planet although it covers a period outside the scope of this book, which, at least in theory, only continues until the emergence of human intelligence (equated with the dawn of humanity, homo sapiens sapiens, or modern humans, depending on how readers see it), which I propose to have occurred from 75,000 to sometime around 60,000 years ago. Dates are really approximations, in deference to logic rather than to fossil evidence, the dating of which invariably raises doubts and provokes discussions for decades without a consensus being reached. By the way, as I mentioned, at the end of 2013, several news portals reported a new carbon radio dating method called ultrafiltration, which removes the most recent carbon molecules that may have contaminated some fossil bones, making them look younger than they really are. Applied to two of Spain's 11 Neanderthal sites, it showed dates of more than 40,000 years when compared with 32,000 years using the previous method, which, if confirmed in other European Neanderthal sites, may make contact between modern humans and Neanderthals impossible unless sites are discovered in Europe with older dating for modern humans as well.

I propose, then, that about 60,000 years ago, modern humans began to migrate from the extreme south to the interior of the continent, and after a period of time sufficient to provoke a marked genetic divergence, they began to migrate again from southern Africa, now along the eastern seaboard, to populate the entire continent, when at least one group reached Asia through the Middle East, reaching China probably 55,000 years ago and Oceania 50,000 years ago. In reverse order, as I suggested a little earlier, they began to migrate out of Africa and only then inland. At this juncture, I think it is appropriate to set out in Figure 16 a map similar to the one shown in Chapter 1, Figure 1, where I synthesize the view of science with regard to the migrations of modern humans so that readers can have a clearer view that my ideas fit well into this picture, except for the date of 70000 years ago for the departure to Asia as this would entail a shortening of the development time of human intelligence, which I calculated to be 15000 years ago (75 - 60 = 15), particularly since, for practical reasons, I cite throughout the book

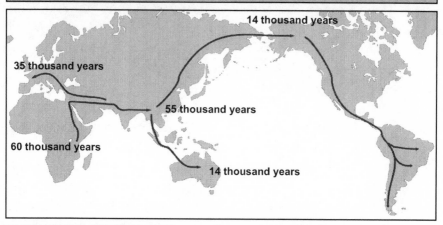

Figure 16

the eruption of Toba as having occurred 75,000 years ago, whereas most of the texts I read estimate it to have occurred 73000–78000 years ago, which would be an easier timeframe for the occurrence of events in the extreme south of Africa. In any case, as I mentioned earlier, dating always provokes doubts and discussions. That date of 70,000 years ago may be 80,000, but it may also be 60,000, which is a possibility that I admit and place on the map. Thus, the map is practically the same as the other except that the date in the first map is 60,000 years ago, which is replaced with 60,000 years ago, and for the arrival in South Asia that in the first map is 60,000 years ago, which is changed to 55,000 years ago, just to make it compatible, that is, without the slightest purpose of being accurate.

Here I suggest that a handful of modern humans, who migrated to Oceania, arrived at Flores Island, Indonesia, and settled there after suffering a phenomenon called island dwarfism, thus tracing back the origins of the so-called *Homo floresiensis*, whose fossils, dating to 12,000 years ago, were found by Australian researchers in 2005. I, therefore, agree with scientists who claim that *Homo floresiensis* was a *Homo sapiens sapiens* as opposed to other scientists who believe that it was a direct descendant of *Homo erectus* (WONG, Kate. *The Smallest of Humans, Scientific American Brazil*, Special Issue, No. 17).

Thirty-five thousand years ago, however, migrations of modern humans from Asia to Europe would have taken place, which may have led to the encounter between the two apparently existing hominids in the world, Neanderthals and modern humans themselves, a fact that may have caused the extinction of the former, a thesis that is threatened by the new dating made in the Neanderthal sites of Spain as I said a bit earlier. In any case, even with the unreliability of dating in the period from 60,000 to 30,000 years ago in Europe, it is true that Neanderthals lived alone on the continent until 60,000 years ago and that after 30,000 years they disappeared (KLEIN, Richard G., BLAKE, Edgar. *The Dawn of Human Culture*, p. 174).

Migration to the Americas caused even more divergence. The Clovis Theory, which argues that there was a migration through the Bering Straits about 14,000 years ago, is the most accepted (DIAMOND, Jared. *The Third Chimpanzee*, p. 371) or at least the oldest. It is based on spearheads and arrows found near the city of Clovis in New Mexico, United States, dated to be 12,000 years old. Several scientists, such as Brazilian archaeologist Niède Guidon, who studies the Serra da Capivara National Park in São Raimundo Nonato, Piauí (the state where I reside), Brazil, which has the largest archaeological site of prehistoric paintings in the world; biologist and anthropologist Walter Neves, who directs studies of fossils found in the region of Lagoa Santa, Minas Gerais, in which the skull called "Luzia" stands out; and Professor Michael Waters of Texas A & M University, who discovered spearheads at the Debra L. Friedkin archaeological site in Texas, United States, dating back 15,000 years, question Clovis and advocate at least two major migrations, an older one, with humans of physical aspects more similar to the Australian Amerindians, and another with humans of physical aspects more similar to the peoples of extreme East Asia.

Specialists from around the world have dated the Serra da Capivara finds, with results even supporting dating them back to 50,000 years ago. However, archaeologist Niède Guidon believes that they can date back 100,000 years. I do not have the technical authority to offer an opinion on dating; however, I acknowledge that the Toba side of my theory would be undermined if modern humans were proven to have existed 100000 years ago in South America. After all, according to my ideas, humanity would not have even existed at that time as I propose the emergence of human intelligence around 75000 years ago, with the mega explosion of the Toba volcano. To be compatible, a date error of approximately 40,000 years would have to be accepted, which would be inconceivable. A dating of 50,000 years, however, would be compatible as, if humans arrived in Australia by that time, they might also have arrived in America around the same time via the Bering Strait in ice ages, when the level of the oceans declined extraordinarily. This agrees even with the conclusions of Walter Neves, who suggests that the American continent was colonized by at least two different biological populations that succeeded each other in time. Neves, however, admits that even with the fossils of Lagoa Santa, in Minas Gerais, Brazil, the earliest date for the first migration is 14,000 years ago. If the Toba strand is removed, however, my theory may even be compatible with modern humans dating to 100,000 years ago anywhere on the planet.

27

The possibility of proof

As usually happens with every new theory, the first question that arises is whether there is a way for it to be proved. Since evolutionary theories, or rather theories related to the evolution of living beings, deal with facts that occurred thousands of years ago and even millions and billions of years ago, they are, in general, practically impossible to prove. To be considered, or at least to be quoted by consecrated authors, is certainly the first step toward being accepted in the future as part of the concepts generally mentioned in evolutionary studies. It is a long road, to reach the end of which most authors of these theories do not survive, and even if some do, the end of the road is not always what they have dreamed of. This way, there is always the risk that the idea will be destroyed by a new fact, particularly by fossil finds. If my theory can remain viable for a few decades, this is most likely to happen to me: I will not see the glory; however, I will escape the weight of failure. Indeed, proving the central theory of this book—that of the emergence of extraordinary and unique human intelligence—as with the great majority of evolutionary theories, proves to be unworkable.

However, one might at least test whether prematurity, occurring with a genetically close species of modern humans in a manner that is similar to what I say happened to a group of hominids in the extreme south of Africa, may cause an increase in their cognitive ability. For this, a situation similar to the one I state in Chapter 7, which provoked the appearance of human intelligence,

225

would be artificially created: a narrowing of the pelvis causes prematurity, which causes babies to be born before their instincts are completely engraved and with more free memories. This provides a lasting environment, which is favorable to the emergence of algorithms able to take advantage of the memory spaces, particularly in a situation of extreme difficulty in life that decisively favors more intelligent animals. Upon synthesizing, two genetic bottlenecks would be artificially created: one aiming to induce premature births at a constant rate of increase and another to select animals with greater cognitive capacity.

Before continuing, let me make it clear that I am not proposing to create a human being from a monkey or any other animal. Also, I am not suggesting that anyone perform this experiment because some of their procedures may confront customs and/or breach legal and/or religious norms of some human communities scattered around the planet, for which I have the greatest respect. I am only reporting—and I think this is my duty toward my readers— that there is a way to demonstrate that the main concept of my Human Intelligence Emergence Theory, prematurity, can significantly influence the degree of increase of the intelligence of animals that are genetically close to modern humans. With this done, the objective of the study would be met and the experiment could be ended. Perhaps, it should be ended because as far as I know, with the randomness of genetic changes, nothing prevents a single change from causing a flood of others capable of pulling a trigger that produces a being more intelligent than modern humans, with absolutely unpredictable consequences. Although improbable, nothing prevents this as I explain at the end of this chapter.

A caveat: this experiment is far more complicated than it appears to be at first glance. The term "complicated" here must be interpreted with much rigor and exaggeration. Therefore, I am now presenting "a synthesis of the summary of a project," if I may so express myself, to emphasize that in the case of a real situation, hundreds of variables would have to be assessed and controlled for. However, merely to demonstrate the possibility of reaching a conclusive result, as I am doing now, there is no need for so much detail.

Initially, I need to explain the problem of time because I propose 15,000 years for the total emergence of human intelligence, a period that makes any test with the technologies we have today impossible. However, I try to show

how this can be overcome by resorting to the work carried out with silver foxes by DK Belyaev and his colleagues, which I quote in Chapter 10. Crossing the meekest animals of each generation in just 20 years, they ended up with foxes that behaved like Border Collies. As I said at the time, genetically tame foxes had black and white fur with white spots on their faces and muzzles, the pointed ears of their ancestors became fallen ears, they acquired a new hormonal balance, and they reproduced all year round instead of doing so in a specific season as their ancestors had done in the wild. Therefore, many modifications can be achieved in a short time when we induce an artificial genetic bottleneck as was done with silver foxes in relation to meekness among humans. As the purpose is not precisely to create an animal with an intelligence that is equal to that of modern humans but only to know whether, in circumstances similar to those that I propose in my theory, there may be a sensible increase in cognition, I imagine that a conclusive result can be reached within a time frame similar to that of Belyaev's test with silver foxes or perhaps in an even shorter time, particularly if we interfere to increase the effects of the genetic bottleneck that favors animals with a higher degree of intelligence.

From here on, some premises may better clarify what might be a real experiment. The first is the choice of the two species to be evaluated because I preferred to select two to gage important characteristics for the acceleration of the process of increase in cognition. In a true experiment, surely this choice should be preceded by a detailed study of several species, particularly in the sense of avoiding characteristics in the subjects that impede the chances of reaching a conclusive result. Thus, for this explanation, I used a few parameters, among which was genetic proximity to modern humans, hands to manipulate objects, gestation time, mating age, childhood duration, and life span. I opted, then, for a kind of nail monkeys and a kind of rats. I remember that the terms "foxes," "nail monkeys," and "rats" cover many species. Thus, in a concrete test, the choice between species of both nail monkeys and rats, if these groups were chosen, would also have to be preceded by a detailed study of the best options.

My choice of nail monkeys was influenced mainly by their genetic proximity to modern humans and their agility in the manipulation of objects. However, what proved to be expedient in the determination of the rats was mainly the small generational lifespan, which, in theory, provides a greater speed in the appearance of changes that could be used in the new environment of induced prematurity. It should be noted that rats, despite having a tiny opposable thumb, have some manual agility, including using their paws to carry food to the mouth. In a real case, this should be observed, both for rat species selection and for the elimination of study mice.

To facilitate readers' understanding, I developed three tables that I show in Figure 17, with some information from modern humans, foxes, nail monkeys, and rats. By the way, the foxes enter this table so that one can compare the actual experience of Belyaev with the study now proposed. Data from foxes, nail monkeys, and rats are averages because the words cover a number of species as I said above.

The calculation that I use for the lifespan of a generation is made using the following formula:

$$G = (g + r) \times 1.1,$$

where **G** is the time of a generation, **g** is the time of a gestation, **r** is the time for the animal to reproduce, and **1.1** is a factor that adds a tolerance of 10% to any mismatch between partners.

In a controlled environment, I believe that this calculation closely matches the real time between a generation and the previous one. A generation in modern humans is often stipulated as 20 years. This is perhaps because culturally 18 years is still considered to be an ideal age for weddings. My estimate, however, is 15.13 years because in this work, the animals will mate as soon as they reach the age of procreation, not facing the natural inclemencies of life in the wild, which often hinder encounters.

I confess that I was tempted to use marmosets, a denomination that covers a variety of very small monkeys, instead of nail monkeys because they have genetic proximity to modern humans that is similar to that of nail monkeys and their generation takes less than half of the time taken in nail monkeys generation. By the way, their gestation is 153 days, they reach reproductive maturity in 16 months, their childhood lasts around 6 months, and they can live for up to 20 years. Their generation, according to my formula, would take 1.93 years, that is, as I said a little earlier, less than half of that of nail monkeys. This example provides a clearer way for the reader to understand the complications of choices in work of this kind.

Characteristics addressed			
	Modern H.	Foxes	Nail monkeys	Rats
Gestation	270 days	52 days	170 days	20 days
Reproduction	13 years	10 months	4 years	50 days
Infancy	6 years	8 months	1 year	20 days
Life	75 years	10 years	40 years	2 years

Characteristics addressed (in days)			
	Modern H.	Foxes	Nail monkeys	Rats
Gestation	270	52	170	20
Reproduction	4680	300	1440	50
Infancy	2160	240	360	20
Life	27.000	3.600	14.400	720

Generation			
	Modern H.	Foxes	Nail monkeys	Rats
Days	5.445,00	387,20	1.771,00	77,00
Months	181,50	12,91	59,03	2,57
Years	15,13	1,08	4,92	0,21

Figure 17

With the species designated, I try to synthesize the procedures, particularizing them and trying to adapt them, when necessary, to the selected animals.

Two environments would be created, each appropriate to a species, where a certain number of individuals, defined in the planning of the study, would be observed by separate scientists, including physicians, veterinarians, zoologists, anthropologists, engineers, paleoanthropologists, biologists, and others who may be required. It is worth emphasizing that each environment can have a different number of animals according to the peculiarities of each species. These numbers could vary according to the needs observed in the course of the experiment, which would always aim for prolonged subsistence, possibly for dozens of years.

In the case of nail monkeys, considering that in nature they live in small groups but in large areas, the ideal environment would be a small sea island to ensure a space similar to the natural one and for isolation. For rats, it should be a smaller environment, perhaps spanning 2–3 hectares, and could be on the mainland but would have to be hermetically sealed, also to ensure a natural environment and isolation. The two environments should be located in regions with very cold climate, probably above the 50th parallel north or below the 50th parallel south, with the aim of emulating a climatic conjuncture that is as close as possible to the situation of the volcanic winter climate faced by hominids in southern Africa 75,000 years ago as I say in my theory—and not just for that reason as we shall see. Quite cold climates can provide at least seemingly advantageous situations for increases in the degree of cognition. First, they conserve food for much longer, making the hunting of larger animals much more profitable. Second, they oblige animals, particularly those accustomed to more temperate environments, to seek shelter. The combination of these two factors favors behaviors related to the increase in the degree of cognition such as 1) meditation—because extreme cold obliges them to protect themselves for days together without obligatory functions in shelters, which are usually caves, thus culminating in a situation conducive to reflective thinking—and 2) socialization—because normally, shelters are rare, due to which a large number of individuals are forced to live in a restricted space, which in turn necessitates more complex collective relationships. Thus, during winters, a greater quantity of perishable food should be made available to the animals to encourage storage. These assumptions restrict the choices of environments to Northern Europe, Asia, and North America, and the extreme south of South America. New Zealand could be assessed due to its proximity to one of the borders considered, the 50th parallel.

Once the numbers of animals has been defined and the environments established, the experiment begins, mainly through two actions:

1 - Artificially forcing the reduction of gestation time so that for a certain period, it progressively falls to a proportion similar to that which my theory maintains—that the gestation interval of modern humans fell in relation to that of the Neanderthals, which was 13, and remained at 9 months;

2 - Producing, also artificially, situations that provoke the emergence of an environment in which only the most intelligent beings can survive.

For the success of these two procedures, various attitudes will be needed to accelerate the process and to avoid side effects that clash with religious and cultural principles. Here are a few so that readers may have a better view of the possibilities of the work being done successfully:

a) Defining, as precisely as possible, the gestation time of the analyzed species;

b) Identifying the beginning of pregnancies as accurately as possible through a methodology that can be repeated throughout the process;

(c) Researching and choosing a feasible method for inducing premature births for a certain number of generations so that over a period of time, it progressively falls to a proportion similar to that which my theory holds for modern humans when compared with the Neanderthals;

d) Placing cognitive evaluation tests in the form of difficulties in obtaining food within the reclusive environment of the animals so that the most intelligent can be identified by direct observation and/or by the state of nutrition, which must be absolutely controlled to avoid deaths;

(e) Setting a period for the removal of animals so that a population in both numbers and sex is maintained at all times to ensure the continuity of work in the reclusive environment;

f) Removing the less intelligent animals, leaving in the reclusive environment a population according to the criteria of the previous item.

Regarding item (c), except for some technical solution of which I am not aware, cesarean deliveries may be used. With respect to item (d), the difficulties in obtaining food must be varied enough to require even cooperation among individuals. Regarding item (e), it should be noted that the withdrawal of animals that cannot solve the difficulties of obtaining food is intended to prevent the animals from dying from starvation as usually occurs in genetic bottleneck environments.

At this point, to have an understanding of the real goals of the study, readers need to know what I really expect to happen when I speak of manifestations that demonstrate an increase in the cognitive ability of a being genetically close to modern humans. In general, science has related human intelligence to the emergence of language. However, since I consider that each animal has its language and that the complexity of this language depends on the degree of its intelligence, it is meaningless to use this as a parameter. I would run into the same problem that I faced in distinguishing the intelligence of modern humans from the intelligence of other animals and would not know precisely the frontiers within which the language of the animals under study could be considered within the limits of human language, a finding that would make it possible to end the experiment. For this reason, I prefer to consider reaching the objective of the experiment when one of the animals is able to make a drawing representative of something important for its existence, thus demonstrating, without a shadow of doubt, the ability to think symbolically. Of course, both language and other higher-order manifestations associated with cognition, such as the ability to store data, process data, make the right decision, create new advantageous actions, repeat advantageous actions, avoid harmful actions, learn how to proceed, adapt to new environments, plan new beneficial actions, and comply with plans, would be noted, studied, and considered during the test and in subsequent analyses of the results. However, the mark that would define the beginning of the emergence of an intelligence similar to that of modern humans would be the undoubted confirmation that the animal was able to make a drawing representing something important for its life as modern humans did, for example, in the caves of Chauvet, France, about 30,000 years ago, and in the cave of Altamira, Spain, between 16,000 and 32,000 years ago. It is not necessary for the drawing to be of the same quality because the drawings of Chauvet and Altamira are really impressive considering the periods in which they were dated to have been made. So much so that they came to be considered frauds when they were divulged.

After all, drawing really seems to be a remarkable, relevant, and decisive event for the emergence of humanity. The planet Earth existed 4.5 billion years ago, remote life existed 3.6 billion years ago, fish inhabited the seas 500 million years ago, dinosaurs dominated the planet between 230 and 75 million years ago, mammals appeared about 200 million years ago, the earliest hominids began to roam the African continent 7 million years ago, and the fossil record has never given the faintest hint that there ever was a being with intelligence similar to that of modern humans up to 40000 years ago. From that date, however, fossil records begin to show indubitable marks of stone designs. Then, the authors of these rudimentary drawings began to produce more elaborate drawings, sculptures, and adornments; they began to construct dwellings, which were initially simple, and then had several floors, eventually arriving at buildings with hundreds of storeys; they invented agriculture, commerce, writing, science, cities, and countries; they constructed means of locomotion such as carriages, automobiles, trains, airplanes, and spaceships; they conquered the entire planet, arrived on the moon, and are currently sending probes to other planets and comets in the solar system. Therefore, I have no doubt that if an animal were to produce a picture of something important for its survival, the experiment could be ended because it is clear that it would at least be starting a cognitive increase similar to what happened with modern humans. However, if even one of the animals can make a drawing but doubts remain, the study could continue until unquestionable evidence is detected.

The reasoning of the previous paragraph points to the need to provide the necessary means for the animals analyzed to produce a design if at some point in the study, they acquire a degree of intelligence sufficient to do so, to avoid the occurrence of what presumably befell the first intelligent hominids, who had difficulty in finding materials to produce their first drawings. This way, a place would be chosen within each environment, where something similar to wax chalk would be scattered on the floor near smooth, rock-like surfaces throughout the entire period of experience.

One last problem remains: what do we do with the remaining animals after the end of the research if a significant increase in intelligence is observed? In other words, what do we do with the animals of the last generations who have cognition far superior to the cognition of their species? It might be dangerous to mix them in the wild with other animals of the same species because they may have different neurological modifications and could spread these to the wild population in an uncontrolled environment with unpredictable consequences as I pointed out earlier when commenting about the end of the experiment.

However, the most logical reasoning indicates that there is no danger in the insertion of these animals into nature as I also mentioned at that moment. If, at the end of the experiment, an animal with a higher cognition is developed, it is expected that the offspring of this animal will only become as intelligent as the father/mother if they are also premature because the new algorithms will need free memories to produce greater intelligence. Seen from another perspective, as prematurity is artificial, when these animals are placed in the wild, their children will be born again at the right time, with their memories already filled with instincts that will not allow the full use of these spaces by new algorithms. Prematurity, according to my theory, is therefore essential for the complete increase in the degree of intelligence. Once deleted from the process, new algorithms will be lost in later generations. This argument is also valid for animals to be removed periodically as there is a risk that they may also have cognitive abilities modified, albeit minimally; however, all this is in theory. In a specific case, important decisions, such as those involving environmental risks, should be made by the project managers in due course considering all the information obtained in the course of the study.

I now close explanations on the Human Intelligence Emergence Theory, and go on, in Part V, to comment on and republish Raffle Theory, a source of inspiration for much of the reasoning used in this book. By the way, I note once again that Raffle Theory is completely independent of the Human Intelligence Emergence Theory and that its publication is intended only to show readers from where the first reasoning came that led me to the ideas that I present in this book.

Part V

Raffle Theory

28

The unpredictability of it all

When I began to plan this book, the intention was to write only about my theory for the emergence of human intelligence and only on how, technically, the facts would have occurred, starting with an urgency in narrowing the pelvis, which is followed by prematurity and the emergence of new spaces of free memories in the brain, and which ends with the use of these free memories to save data as well as to host new algorithms capable of using these spaces to solve complex problems. I would not even try to define a more precise time frame for events. So as not to run the risk of error, I would broaden as much as possible this space of time for the emergence of human intelligence, perhaps between 2 million and 40000 years ago. I wanted to write an uncommitted text to see whether my ideas might have the slightest chance of being right.

However, when I began to write, I saw that I was not able to conceal the origin of the original thought that somehow led me to the rationale of the main theory developed in this book. I felt that it was not right to leave unsaid how it all began. It was as if I were hiding the part most difficult to accept, Raffle Theory, in order not to disrupt a more complacent view of the part that is easier to understand, the Human Intelligence Emergence Theory. It seemed as if I was making a political decision so as not to be confused with a lunatic who raves that "we are the most perfect beings in the universe and that we have been created in the likeness of God" but has decided nothing and is only deceived into thinking that they are in charge when they only obey all that a cold and calculating brain tells them to do. It was there that I came up with the idea of a commitment to sincerity, which, in addition to facilitating my intimacy with readers, required me to tell the true story: that all my ideas emerged, whether right or wrong, or logical or illogical, or absurd or not, from an initial idea that the human brain makes its decisions through a process similar to that of a raffle, which is based on an extraordinary database of graduated preferences. I know that this connection is complicated, but it happened like this, as I try to demonstrate here in Part V.

I then decided to place at the end of this book an exact copy of the article that I posted on the Internet, summarizing my theory to explain the formation of human thinking on the basis of draws. But then, I discovered two things that I consider to be necessary for a better understanding of the proposal. First, I would need to adapt the article to the style of the book. Second, it would be necessary to write an initial piece of text, which would prepare readers to understand the theory, and to provide a brief explanation of the theoretical model so that the reader can see where the raffle process, which defines everything that we think, fits. In any case, with this preparation, I realized why many of those who read the text on the Internet did not understand it. It was not the fault of the readers. The fault was mine. For a minimum understanding, more data was needed, as well as clearer explanation, which I hope to achieve now with this chapter.

To follow the explanations, a minimum understanding of computer science is sufficient. However, readers should note that when I speak of minimal understanding, I mean this literally. That is right. It is sufficient for a reader to be familiar with database tables and some arguments and functions, such as "if this is equal to that, then it will assume a certain value." To know now, for example, that if there is a pay table with three items—name, date of birth, and salary—a query can be made, for example, seeking the name of all employees who earn more than $1,000 and who have their birthdays in February. Knowing that a database can sort data such as that when asked whether a person would look left or right, the program will cast lots giving a 40% chance of looking one way and a 60% chance of looking the other way. I suppose that simple things like these are enough.

It is interesting to recall that one theory may arise because of the other, but either one can easily survive if the other is wrong. That is, they are independent. More precisely, Raffle Theory served as an inspiration for the Human Intelligence Emergence Theory, the explanation of which is the purpose of this book. This little argument and the publication of the article on Raffle Theory only seek to show the logical bases that led me to the first reasoning of the Human Intelligence Emergence Theory. Perhaps one day I will write a detailed book on the processes that I believe determine the brain's decisions regarding living things.

The presentation of the theoretical model requires a small chronological explanation. When I conceived the idea of the raffle, I did not think of splitting the brain into parts nor did I even know where information is processed. I did not consider myself to be prepared to deal with the matter this way. I did not have the technical knowledge to do so and had linguistic and financial difficulties that restricted my access to more specific readings. However, since I make an analogy with computers, I decided to suggest an algorithm that explained human decisions, leaving the study of hardware to those more specialized, that is, in the functioning of the parts of the body involved in the process. I would concentrate only on the logical process.

The best comparison for the idea of a raffle is perhaps a command routine that can be used by a multitude of software; evolving the recording, storage, and retrieval of information; and aiming at making decisions, both in terms of decision making and in more mundane ways, for example, deciding whether to leave or stay at home and whether to carefully remember or discount a particular fact.

Although the theory is general for living beings, I develop an explanatory model using human beings as an example to facilitate understanding. In the original article, I use an experiment with mice.

Imagine the human mind containing two main parts, one that calculates and decides everything, which I will call **Decision**, and another that contains all the information more related to the individuality of a human being, which I will call **Consciousness**. I do not know whether the terms that I have chosen are the most appropriate for the functions that I imagine they carry out; however, as a didactic device, I think it helps to explain the theory, in particular when we know that one cannot be as precise or adjusted as one would like to be, always leaving some room for refinement for someone better able to do so. A word of caution: I am simplifying this to aid assimilation.

In reality, the brain performs numerous functions simultaneously with a complexity that humans do not have the ability to evaluate. Therefore, the reductionism that I use seeks only to facilitate understanding.

Therefore, I think it is essential in the understanding of Raffle Theory that one part of the brain decides what to do and informs the other part of the brain that it, the other part, has made the decision. Using my explanatory model, **Decision** decides and **Consciousness**, fooled by Decision, thinks it has decided. This, in practice, does make people think that they are deciding everything because consciousness is a kind of play of the whole life of an individual. It is truly appalling to admit this while having the complete sensation that we are making all the decisions. Apparently, it seems to point toward determinism, but this is not what happens because despite the use of graduated preferences, what finally makes the decision is a raffle. And what happens, contrary to what appears to happen in principle, is **unpredictability**, which, if it exists in living beings, may indicate that it also exists throughout the universe, suggesting the **unpredictability of everything**.

It is difficult to become accustomed to rules that wound common sense, and adapting myself to the extravagant idea that I have an independent decision-making system took a while. However, convinced that I was right, I read everything I could on the subject until, by the end of 2009, I had acquired a collection of publications on the human brain and learned of an experiment carried out in the early 1980s by the neuroscientist Benjamin Libet, who arrived at the same conclusion that I arrived at by logical reasoning; that is, the unconscious brain decides everything and consciousness only receives the information later.

"Conscious decisions (Benjamin Libet)

A series of ingenious experiments by US neuroscientist Benjamin Libet in the early 1980s demonstrated that what we think are conscious "decisions" to act upon are actually just a recognition of what the unconscious brain is already doing. Libet's experiments have profound philosophical implications because on the face of it, the results suggest that we do not have a conscious choice about what we do and therefore cannot consider ourselves to have free will." (*)

---

* MORAES, Alberto Parahyba Quantin de. *O livro do cérebro.* V. 1: funções e anatomia. Tradução: Peter Frances. São Paulo: Duetto, 2009.

Benjamin Libet was born on April 12, 1916, in Chicago, Illinois, and died on July 23, 2007, in Davis, California. In 2003, he received the Virtual Nobel Prize in Psychology from the University of Klagenfurt, "for his pioneering achievements in the experimental investigation of consciousness, initiation of action, and free will"—a weighty name, indeed.

I was surprised: I had deduced, around 2006, what a great scientist had proven through experiment 26 years earlier. And I conducted no experiments: I deduced. What is more, I believe in the result of Libet's experiments, and Libet himself did not. He proposed a solution that maintains free will. Further, he did not explain how the brain decides. One of Libet's experiments clearly shows that an action (for example, of picking up a pen) occurs first and only then does the information stream into consciousness. This way, the decision to perform an action certainly occurs without the participation of the conscience. I knew that this did not mean that I was right even though it has been proven in the experiments of several other scientists. Even Libet, as I said a little earlier, did not exactly agree with the results of his experiment and argued that even with the decision occurring too early to be triggered by consciousness, there would still be a time window, lasting around 100 milliseconds, in which consciousness could veto action. His results were so contrary to common sense that the scientist himself sought an answer that established reason and maintained free will.

Perhaps because I came to this conclusion by logical reasoning, I reaffirm my argument that the brain really decides everything and that consciousness is deceived into thinking that it is deciding something even while acknowledging that this is a paradox. Somehow, perhaps because I have lived with the theory for some years, I no longer see the process as a contradiction, and I have become accustomed to analyzing people and myself as if the theory were true, and at least in relation to events and decisions, everything seems to me to be happening in deference to Raffle Theory, which uses the trends recorded in the databases that we have scattered in our body and that are probably centralized in the brain. A good idea for the reader to become more familiar with Raffle Theory is to forget about consciousness as we feel it and to imagine the human brain as a machine for producing answers, which eventually has to draw lots whenever the parameters are insufficient to set or even to reflect that testing other options is beneficial to the species, however unusual, even absurd, that these options may seem.

One point, however, needs to be clarified: can a computer system carry out a raffle? As far as my understanding goes, the answer is no. And that is not good for Raffle Theory. However, it is true: closed computer systems cannot hold a raffle. They simulate a draw based on prerecorded random numeric sequences as I have studied in various texts on the Internet. Seen in this light, Raffle Theory could lead to determinism as the draws would not be realized but only simulated on the basis of a numerical sequence that humans would have recorded somewhere. The reasoning is correct; however, I have my doubts. Although I agree that a closed computer system cannot hold a draw, I suggest that in some way, the human mind cannot understand that nature has figured out a way to hold a raffle. Or perhaps the system is inherent in nature, which is something similar to the Theory of Uncertainty at the atomic level. By this understanding, the idea of draws may even be the explanation for the random mutations that occur in living things or who knows, the basis of all evolution and even the whole universe.

In the next chapter, I present the article published on the Internet in 2006, edited for use in this book. I confidently hope that it will now be better understood.

29

Primordial logic:
Raffle Theory

(Written in October 2005 and originally published on the Internet on April 18, 2006. The text is basically the same as the Internet version, with only a few corrections, to retain the style of the book.)

Note: I maintain at the end of the explanation the original references of the article that I published on the Internet.

The special issue of *Life – Mind and Brain*, N° 2, published in 2005, under license from *Scientific American*, Inc., which is dedicated to memory, is entitled *Memory – the physiological, neural, and psychic bases of the memory file*, and has several articles signed by renowned scientists, deals with innumerable issues related to memory and the functioning of the brain (or the nervous system in general) but does not talk about how thought works or how the mechanism of remembering works—in short, nothing is proposed about the logic, the process that determines the activities of the brain, or more broadly, the whole system that processes the decisions of living beings. It calmly discusses the tendency to remember or to forget; however, no attempt is made at any time to explain the logical process of thought, remembrance, and forgetfulness as if thinking that remembering and forgetting were definitive and final facts. The purpose of the theory that I put forth is precisely to explain the logic of the process of the functioning of the nervous system in general, including the mind, thought, memory, consciousness, character, and personality.

Scientific American magazine, founded in 1845, is one of the most important publications in the world that treats the progress of science and technology. When it comes out with a special edition, it presents the latest information on a subject. Therefore, it is normal to imagine that there is still no explanation or at least no acceptable explanation about the logic of the brain's functioning process—and it is true as I have seen in several works of respected scientists.

In her article What is the Mind?, published on the website of Brain & Mind magazine, www.cerebromente.org.br, Dr. Sílvia Helena Cardoso (1), to demonstrate our lack of knowledge about the mind, wrote the following:

"It is impressive to note that even after several centuries of philosophical reflections, arduous dedication to brain research, and remarkable advances in the field of neurosciences, the concept of the mind still remains obscure, controversial, and impossible to define within the limits of our language."

Vilayanur S. Ramachandran (2), in his article The Future of Brain Research, on the same site, was even more detailed in assessing how little we know about the process of organ functioning and the doubts that science has today about the subject:

"The scientific revolution completely changed our view of the universe and our place in it, but the best of all (or worse!) is yet to come. It is ironic that we have so much detailed knowledge about almost everything that exists in the universe on every conceivable scale—the solar system, distant galaxies, black holes, atoms, molecules, string theory, DNA, heredity, the mechanisms of life, etc.— but we know almost nothing about the organ that made all these discoveries.

Certainly, our knowledge of brain functioning today remains as primitive as our knowledge of the rest of the human body a century or two ago (for example, about the liver, spleen, or pancreas). A century ago, we had only a few vague notions of the liver having "something to do with digestion", now we know that it has more than 30 metabolic functions, each of which is understood in all its intricate details. However, despite the accumulation of vast amounts of factual knowledge about the brain (about 10,000 papers are presented each year at meetings of the Society for Neuroscience in the US), most BASIC questions about our minds remain unanswered. What is will? What is 'I'? Why do I feel like a unique person, who resists through time and space? What is consciousness? Why do we laugh, making head movements and rhythmic vocalizations in certain situations? Why do we cry (which could also be put as follows: why does this salty liquid run down my cheeks?) Why do we dance? What is the meaning of art? Why does music exist? Why do we dream? Why do we need sleep?

However, I will take the risk of predicting that all of this will change in the next 100 years. By asking the right questions and doing the right kinds of experiments, we can begin to answer these questions so well that until now have been the concern of philosophers."

As the reader can observe, little is actually known about matters connected with our minds, and to answer most of these questions, it is necessary to understand the logic of the process of the brain's functioning, thus suggesting a theoretical model explaining how these phenomena occur. This is what I intend to do. If this book has any editorial success, I know that this theory will displease scientists and the religious. To admit that everything we are and do depends on the fatality of a lottery governed by a set of probabilities is really shocking, both for science, with regard to the concepts of the mind, thought, memory, consciousness, character, and personality, and for religion, with regard to the faith that most humans have, which is almost always based on the belief that every living human body has a supernatural spirit or soul that thinks and makes decisions without the necessity of a mechanism for it.

I believe that the mind can only remember information that the brain has recorded completely down to the smallest detail. So, if we forget and then remember, somehow, nature has forged a system in which the mind does not remember information that the brain has kept in full detail. Another important factor is the unpredictability of memory. That is, the mind can remember information at a given time and then forget, only to later remember again. The logical process that I have encountered to solve both situations, with the brain having complete information but the mind not having access to it at any given time, only to retrieve it at another time, is that the brain marks the information with a probability and carries out a raffle in deference to this probability when the mind requests for the piece of information. As the reader can observe, to support this logical process, it is necessary to accept the existence of a memory unit or something similar with various pieces of information that allow its identification by a computational system, such as the probability of access, image, number, smell, taste, touch, and even indexes that determine correlation with other memory units.

To demonstrate the basic principle of this theory, I will use a proven experience, which explains the widely accepted phenomenon of the tendency toward alternation, which is explained in the article Evolutive from the Records of Georges Chapouthier, Director of the National Center for Scientific Research (CNRS) in Paris, in the magazine Viver - Mente e Cérebro, n° 2, special edition on the brain, already mentioned at the beginning of this work.

"The paradigms that evaluate the phenomenon of alternation are based on the fact that, if an animal is placed before two choices and several times chooses one, it will have a tendency, statistically, to opt for the other. A mouse in a T-maze, which has repeatedly chosen the left side, will tend to opt for the right side even if there is no particular attraction influencing the choice. The preference for the "other side" obviously supposes that the animal memorized the first side on which it entered. By modifying the nature of the proposed choices, we can adapt this task to all animal groups."

The author says that if the animal comes to have a preference for one side, which is a preference that it did not have previously, it is because it memorized the first side that it entered. OK. Now the question is to know how it memorized and how it uses that memory so that when it sees itself again in the face of the problem, it tends to choose the opposite side. The reader should note that this is only a tendency and not a certainty. That is, it may eventually choose the same side although there is a proven tendency to choose the other side (see the illustration in Figure 18).

I propose that the brain of the rat, when it faces a new situation, in which it must choose between two absolutely equal alternatives, somehow promotes a raffle in which the two choices have a 50% chance of winning. After all, there is nothing that points toward choosing one side, but a decision is made, which is an occurrence that needs to be explained. The draw, from the perspective of those who analyze the fact that there was a decision, is the explanation. The

PHENOMENON OF THE TREND OF ALTERNATION

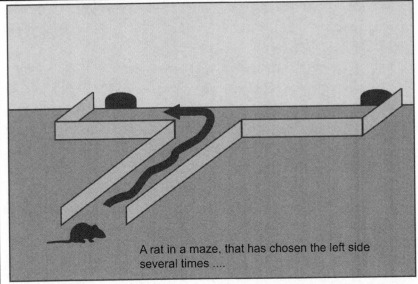

A rat in a maze, that has chosen the left side several times ....

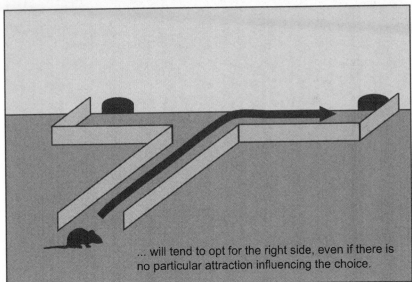

... will tend to opt for the right side, even if there is no particular attraction influencing the choice.

Considering the human brain as a problem-solving machine, something like a computer, I reason that its tendency to choose a path is only explained by the fact that it chooses at random, assuming a greater probability of one choice over another.

Figure 18

draw, from the perspective of the system, if this is how I can express myself, is the resource to answer the question. Once the raffle has been made and where the mouse will go has been decided (to the left, for example), the brain marks that information so that the next time it is accessed, the chances of choosing the right side increase, say, to 60%.

I emphasize that these values have the sole purpose of explaining the process and that the use of percentage proportional comparisons is only to facilitate understanding. I cannot even imagine how the brain processes a single word; hence, I have no way of proposing how a raffle really works.

Let us continue the reasoning. If the rule were that the rat's brain chose a side simply on the basis of the information that it had chosen the left side the first time around, it would go to the right side, obeying alternation. However, that is not the case. It is not a case of pure logic wherein if you went to one side, now you have to go to the other side. It is a trend: if you went to one side, you now have a greater tendency to go to the other. Then, the issue is resolved by only one draw having more probability toward one side, which is the side opposite to the one previously chosen. Once again placed before the problem, the brain of the mouse makes a new draw—only this time, it is more likely (60%) to choose the right side. This way, even with larger chances of the right side being chosen, the choice may be the left side. Well, if the left side is chosen again, the rat's brain marks the information, such that the next time it is accessed, the chances of choosing the right side increase even more, let us say, to 65%. If, however, the right side is chosen, the rat's brain returns to a 50% chance for each side as the choice was once to one side and then to the other.

Does this model match experience? I think so. Does it explain the phenomenon? Ditto. Does it apply to thought in general and ultimately to all the functions of the brain? I believe that it does for most functions.

After much thought on the subject, I have come to the conclusion that it applies not only to most brain functions but also to many other functions of living organisms. However, this is a subject that I intend to deal with at another time, probably when the development of the theory is close to completion.

I propose, therefore, that the brain functions in a continuous process of drawing lots, using programmable probabilities, that decide what I, what the reader, and what all living beings will do in a few moments.

However, let us see how the model works for the most common brain process, which is to record, remember, and most importantly, forget, information. I am not talking about immediately accessing information to make a decision as in the experience of the alternation tendency phenomenon. I am talking about trying to remember information that an individual knew at some point with clarity and without doubt.

Let us suppose that a person, whom we will call A, knows another, whom we will call B. They talk, the conversation turns to birthdates, and B tells A their birthdate. If someone asked this information of A soon thereafter, they will likely remember. At night, if someone asked A whether they knew B's birthdate, A is very likely to know. The same is the case with the next day. However, if the same question is asked of A many days later, A certainly will not know how to respond, and of course, they will say that they forgot. But how did they forget? Is the information lost somewhere in the brain? Was the information deleted?

Let us move on. If A keeps trying to remember B's birthdate for a long time, A may suddenly succeed and remember—complete with day, month, and year. If the information had been erased, how then did the brain recover it intact? How many times has something similar to this happened to all of us? It happens all the time—when we try to remember what we studied, what we heard on the radio, what we saw on television, and what commitments we made; when we look for an object; etc.—and almost always with results similar to that of the example. But, true, sometimes we never remember the information again. There is, therefore, unpredictability—more specifically, biased unpredictability, the tendency of which depends on certain factors. For example, the more time we spend without accessing information the more difficult it becomes to remember it. The less attention we pay when we learn something the more difficult it is to remember. Why do things occur this way? Why does the brain forget and then remember? Why does someone not remember information they knew yesterday and can remember tomorrow? Why does a student study a particular subject diligently and forget it at the time of the test? Why, on leaving the test, can the person suddenly remember it all? That is exactly what the theory seeks to explain.

At the moment that A learned B's birthdate, A's brain recorded the information with a high percentage of chances of remembering it, say, 99%.

So if, soon after, someone asks A about B's birthdate, A would have hardly forgotten because A's brain would process a 99% draw, for example, of the probability of remembering, that is, to access the contents of that memory, and 1% of not remembering, that is, not to access the contents of that memory. With a probability of 99%, the chances of not remembering would be slight. But that 1% would guarantee the possibility, even if very remote, of forgetting. This does in fact occur with most of us. Sometimes, we just become aware of information, and inexplicably, we can no longer remember it.

However, the theory presupposes that the brain is always organizing and updating the parameters of remembrance of all information and that if information is not accessed for some time, its recall propensity decreases. Thus, over time, without being asked B's birthdate, the probability that A will not remember will always increase. At night, for example, A could have a 90% probability of remembering and a 10% probability of forgetting.

Somehow the brain, over time, without the information being accessed, decreases its likelihood of being remembered. Because A was not asked for the information, the probability of their remembering it was always decreasing. If A were only questioned the next day, the probability of remembering it would be even lower, for example, 80%. If many days pass before the information is asked of A, the probability of recollection might already be so small, say 10%, that A would remember B's birthdate only with difficulty. However, if A kept trying to remember, for every failure, A's brain would increase the recall probability, from 10% to 11% to 12%, and at one point, the probability could become so great that when the brain draws the lot, A remembers the information.

And what would happen next? As soon as A recalled it, the brain would immediately raise the probability of remembering to a level similar to that when A learned B's birthdate for the first time, 99%, and if questioned, then A would hardly forget the information.

Does the theory also explain this phenomenon? Once again, I think so. Of course, there are other factors that influence the process of recording information, remembering it, and forgetting it, particularly when we know the complexity of the brain, which allows it to perform several functions simultaneously. However, I imagine that most of these processes, which occur simultaneously and always by means of a raffle with previously established probabilities, forge what we call the mind, thought, memory, consciousness, character, and personality.

Having verified the logical process that I propose with these two phenomena, those of the alternation tendency and of the recording and retrieval of information, I close in this book the presentation of this theory, which I think may be applicable to other phenomena involved with the brain, such as the mind, thought, memory, consciousness, character, and personality, and even to various processes of the functioning of living organisms such as fecundation, mutation, development, organ degeneration, aging, and death, all of which are related to unpredictability. As they say, that is life—unpredictable. And if this theory is correct, Albert Einstein, probably influenced by his religious background, may not have accurately assessed quantum theory when he said "God doesn't play dice." By the same reasoning, with God existing and with my Raffle Theory being correct, He built life on the basis of a dice game—perhaps the whole universe!

I believe that even Einstein's own intelligence can only be explained by a cognitive system that can choose and follow a path of reasoning that is unusual, unexpected, illogical, absurd, and unacceptable for common sense as was the idea of space time.

A brain that decides only on the basis of the senses, human perception, and the concepts of physics of that time, however much it tried to find solutions, would never consider analyzing a path of reasoning as far off the oft-taken path as the one that Einstein followed to present his most famous theory. There would have to have been a raffle in which the brain would almost always look for answers in deference to scientific precepts and common sense. But there would also have to be the chance, even a minimal one, of trying to come up with an idea that would go against common sense and defy established concepts.

Only with this unpredictability am I able to understand a computational process that produces reasoning similar to that of human intelligence.

Teresina, Piauí, Brazil. May 18, 2006.

Euripedes de Aguiar

(Theory conceived in October 2005)

References in this chapter

(1) **Sílvia Helena Cardoso**, PhD, Psychobiologist, with a master's and doctoral degree from the University of São Paulo and a postdoctoral degree from the University of California, Los Angeles. An associate researcher and invited lecturer at the NIB/UNICAMP. Editor-in-Chief and creator of the Brain & Mind magazine.

(2) **V. S. Ramachandran** - Director of the Center for the Brain and Cognition; professor of the Department of Psychology and Neuroscience Program at the University of California, San Diego; and Associate Professor of Biology at the Salk Institute. Ramachandran was trained as a physician and obtained a doctorate of medicine from Stanley Medical College and subsequently a PhD from Trinity College at Cambridge University where he was appointed as Rouse Ball Senior Researcher. Ramachandran's early research focused on visual perception, but he is best known for his work in neurology. He has received many honors and awards, including a scholarship from All Souls College in Oxford; an honorary doctorate from the University of Connecticut; a gold medal from the Australian National University; the Ariens Kappers Medal from the Royal Academy of Sciences of the Netherlands for outstanding contributions to neuroscience; and a Presidential Lecture Award from the American Academy of Neurology. He is also an associate member of the Institute of Neuroscience at the La Jolla Institute and an associate of the Institute for Advanced Studies in Behavioral Sciences at Stanford.

EPILOGUE

Inspired by the following two films: 1) Ulysses, released in 1954, written by seven professionals, including Ben Hecht and Irwin Shaw, and starring Kirk Douglas and 2) Troy, released in 2004, written by David Benioff and starring Brad Pitt, which are adaptations of the Greek epic poems The Odyssey and The Iliad, respectively, I composed the following text to show how, of course, modern humans, both modern-day and of 3,200 years ago, supposedly the date of the Trojan War, are very afraid of death and somehow always seek a way to deceive it.

The reason

When I still had doubts as to whether to write about my theories, my brain and my conscience tried to resolve the issue. Finally, my brain decided and deceived my conscience by saying:

"Long before you could think, I already knew that these ideas would one day appear. And now, they want to spread themselves around the world. They want you to write a book. If you do not write one, you will find peace and tranquility. You will have a quiet old age, along with your children and your grandchildren, and when you have departed, they will remember you. However, when their children depart and then their children as well, your name will be lost. But if you write the book, glory or failure will be yours. It

253

will be able to narrate your ideas for thousands of years, and the world will remember your name. However, you will never again have peace because your glory and your failure go hand in hand with your destiny."

• • •

Three years later...

We humans are haunted by the immensity of the universe and by the infinity of time. And we are always wondering: will our thoughts echo through the centuries? Will our ideas serve any purpose? Will strangers hear our names long after we are dead? Will they imagine who we are? How bravely did we fight for an idea? How intensely do we reason, seeking explanations for the mysteries of our existence?

This is a science book—one of ideas and proposals. As for explanations of the great mysteries of the universe, just as the Greeks Ulysses and Achilles, I provoked the gods. I overcame many Tersalian warriors; I blinded gigantic cyclopses such as Polyphemus, son of Neptune; I was deceived by witches and goddesses and persecuted by gods such as Apollo and Poseidon. However, I knew how to retreat and tie myself to the mast of the ship and to listen to the song of the mermaids, because I also miss the unknown.

For me, it is a terrible anguish to know how mankind came about. For me, it is extremely stressful to know how human thought is processed. For me, it is an endless martyrdom to be absolutely sure that I do not have the slightest control over what I will do in a few moments. I am relieved, however, that the struggle for knowledge will never be forgotten and that neither will those who participate in it.

If one day my story is told, let them say that I read about giants and about humans that rise and fall as corn in winter, but those names will never perish. Let them say that I lived in the time of Elaine Morgan, advocate and developer of theories. Let them say that I lived in the time of Richard Dawkins. And if the main idea of this book is correct, let them say that I have discovered how mankind came about.

254

Just do not say that I did it with the romantic attempts of a Ulysses. Say that I did it with the vain and selfish purposes of an Achilles, such as that which he demonstrated in speaking to his warriors, instigating them to war when he reached the coast of Troy: "Know what is there, waiting, beyond that beach: immortality. Seize it. It is yours." Or when he said to Hector the prince of Troy, after invading the Temple of Apollo and cutting off the head of its golden statue: "They will speak of this war a thousand years from now. And our names will remain."

For I know that throughout the expanse of eternity, the name of he who discovers how mankind has arisen will last through the ages and will never be forgotten.

Teresina, Piauí, Brazil. April 30, 2013.
Euripedes de Aguiar

REFERENCES

Books

The neanderthal's necklace
ARSUAGA, Juan Luis. *O colar do Neandertal: em busca dos primeiros pensadores*. Translation: André de Oliveira. São Paulo: Globo, 2005, 349p. Título original: El colar del neandertal.

A very short history of the world.
BLAINEY, Geoffrey. *Uma breve história do mundo*. São Paulo: Fundamento Educacional, 2008, 342 p.

A short history of nearly everything
BRYSON, Bill. *Breve história de quase tudo*. Translation: Ivo Korytowski. São Paulo: Companhia das Letras, 2005, 541 p.

CASTELO BRANCO, Anfrísio Neto Lobão. *Mandu Ladino*. 2ª ed. Teresina: edição do autor, 2008, 485 p.

_____. *Manual de Psicologia Médica*. Teresina: Comepi, 1983, 112 p.

Genes, peoples and languages
CAVALLI-SFORZA, Luigi Luca. *Genes, povos e línguas*. Translation: Carlos Afonso Mal Ferrari. São Paulo: Companhia das Letras, 2003, 289 p.

The Origin of Species
DARWIN, Charles. *A origem das espécies*. Translation: John Green. Idealização e coordenação: Martin Claret. São Paulo: Martin Claret, 2008, 639 p.

The ancestor's tale – A pilgrimageto the dawn of life
DAWKINS, Richard. *A grande história da evolução – na trilha dos nossos ancestrais*. Colaboração de Yan Wong. Translation: Laura Teixeira Mota. São Paulo: Companhia das Letras, 2009, 759 p.

Unweaving the rainbow

_____. *Desvendando o arco-íris – Ciência, ilusão e encantamento.* Translation: Rosaura Eichenberg. São Paulo: Companhia das Letras, 2000, 416 p.

The selfish gene

_____. *O gene egoísta.* Translation: Rejane Rubino. São Paulo: Companhia das Letras, 2010, 540 p.

The third chimpanzee

DIAMOND, Jared. *O terceiro chimpanzé – a evolução e o futuro do ser humano.* Translation: Maria Cristina Torquilho Cavalcanti. Rio de Janeiro: Record, 2010, 430 p.

Humans before humanity

FOLEY, Robert. *Os humanos antes da humanidade – uma perspectiva evolucionista.* Translation: Patrícia Zimbres. São Paulo: UNESP, 2003, 294 p.

The Scientifc American day in the life of your brain

HORSTMAN, Judith. *24 horas na vida do seu cérebro – Parte 1.* São Paulo: Scientif American; Duetto, 2010.

Evolution in four dimensions

JABLONKA, Eva e LAMB, Marion J. *Evolução em quatro dimensões - DNA, comportamento e a história da vida.* Illustration: Anna Zeligowski. Translation: Claudio Angelo. São Paulo: Companhia das Letras, 2010, 512 p.

The dawn of human culture

KLEIN, Richard G. e BLAKE, Edgar. *O despertar da cultura – a polêmica teoria sobre a origem da criatividade humana.* Translation: Ana Lúcia Vieira de Andrade. Rio de Janeiro: Jorge Zahar, 2004, 252 p.

MARCONI, Marina de Andrade e PRESOTTO, Zelia Maria Neves. *Antropologia.* 7ª ed. São Paulo: Editora Atlas, 2011, 330 p.

The brain

MORAES, Alberto Parahyba Quartin de. O livro do cérebro. V. 1 – Funções e anatomia. Tradução: Peter Frances. São Paulo: Duetto, 2009.

The aquatic ape hypothesis
MORGAN, Elaine. *A hipótese do símio aquático – uma teoria sobre a evolução humana*. Translation: Simone Ribeiro. Porto: Via Óptima; Oficina Editorial, 2004, 167 p.

The naked ape
MORRIS, Desmond. *O macaco nu: um estudo do animal humano*. 18ª ed. Translation: Hermano Neves. Rio de Janeiro: Record, 2010, 272p.

NEVES, Walter e PILO, Luis Beethoven. *O povo de Luzia - em busca dos primeiros americanos*. 1ª ed. São Paulo: Editora Globo, 2008, 334 p.

Mapping Human History
OLSON, Steve. *A História da Humanidade*. Translation: Ronaldo Sergio de Biasi. Rio de Janeiro: Editora Campus, 2003, 312 p.

Nature via nurture
RIDLEY, Matt. *O que nos faz humanos: genes, natureza e experiência*. 2ª ed. Translation: Ryta Vinagre. Rio de Janeiro: Record, 2008, 399 p.

Catching fire – How Cooking Made Us Human
WRANGHAM, Richard. *Pegando fogo – por que cozinhar nos tornou humanos*. Translation: Maria Luiza X. de A. Borges. Rio de Janeiro: Jorge Zahar, 2010, 226 p.

The evolution of God
WRIGHT, Robert. *A evolução de Deus*. Translation: Flávio Demberg. Revisão técnica: Marcelo Timotheo da Costa. Rio de Janeiro: Record. 2012, 698 p.

Articles

AGÊNCIA NOTICIA DE JORNALISMO CIENTÍFICO. *Ainda na gestação*. Ciência & Vida. Edição Especial, Ano II, n° 6 - Psique. Abr. de 2012

_____. *Comportamentos difusos, cérebro diferentes*. Ciência & Vida. Edição Especial, Ano II, n° 6 - Psique. Abr. de 2012.

B. M. De WAAL, Frans. *Sexo e sociedade entre os bonodos*. Scientific American Brasil. Como nos tornamos humanos. São Paulo, edição especial n° 17, 2005.

BEGUN, David R. *O planeta dos antropóides*. Scientific American Brasil. Edição especial n° 17 – Como nos tornamos humanos. São Paulo, 2005.

BERGE, Cristine. *A longa caminhada*. História viva. *A odisseia dos primeiros humanos*. São Paulo, Ano VI, n° 62, 2008.

CALVIN, William H. *A evolução do pensamento*. Scientific American Brasil. Edição especial n° 17 – *Como nos tornamos humanos*. São Paulo, 2005.

CANN, Rebeca L.; WILSON, Alan C. *A recente gênese africana dos humanos*. Scientific American Brasil. Edição especial n° 2 – *Novo olhar sobre a evolução humana*. São Paulo, nov. 2003.

CHAPOUTHIER, Georges. *Registros evolutivos - memória Biológica*. Viver - mente & cérebro. Edição Especial n° 2 - Memória. São Paulo, 2005.

CHAVAILLON, Jean. *O poder do fogo*. História Viva. Ano VI, n° 62 – A odisseia dos primeiros humanos. São Paulo, 2008.

COPPENS, Yves. *Nossa pequena história no tempo*. História Viva, Ano VI, n° 62 – A odisseia dos primeiros humanos. São Paulo, 2008.

DAMASIO, António R. *Lembrando de quando tudo aconteceu*. Scientific American Brasil. Edição especial nº 40 – *Em busca da consciência*. São Paulo, 2011.

DICKSON, James H.; OEGGL, Klaus; HANDLEY, Linda L. *O homem do gelo*. Scientific American Brasil. Edição especial nº 2 – *Novo olhar sobre a evolução humana*. , São Paulo, nov. 2003.

GENESTE, Jean-Michel. *E o homem recriou o mundo*. História Viva, Ano VI, nº 62 – *A odisseia dos primeiros humanos*. São Paulo, 2008.

HAUSER, Marc. *A origem da mente*. Scientific American Brasil. Edição especial nº 40 – *Em busca da consciência*. São Paulo, 2011.

IKEDA, Patrícia, e HUECK, Karin. *Nascer é um parto*. Superinteressante. Design: Rafael Quick. São Paulo, edição 306, jul. 2012.

KINSLEY, Craig Howard, e LAMBERT, Kelly G. *Sabedoria de mãe*. Scientific American Brasil. Edição especial nº 40 – *Em busca da consciência*. São Paulo, 2011.

LEAKEY, Meave; WALKER, Alan. *Os primeiros fósseis hominídeos da África*. Scientific American Brasil. Edição especial nº 2 – *Novo olhar sobre a evolução humana*. São Paulo, nov. 2003.

LEONARD, William R. *Alimentos e evolução humana*. Scientific American Brasil. Edição especial nº 2 – *Novo olhar sobre a evolução humana*. São Paulo, nov. 2003.

LUMLEY, Henry. *O sentido da vida*. *História Viva*. Ano VI, nº 62 – *A odisseia dos primeiros humanos*. São Paulo, 2008.

M. KINGSLAY, David. *Átomos e caracteres*. Scientific American Brasil. Ano VII, edição especial nº 81 – *A mais poderosa ideia da ciência*. São Paulo, fev. de 2009.

MALDONATO, Mauro; e DELL'ORCO, Sílvia. *Esferas conscientes e inconscientes*. Scientific American Brasil. Edição especial nº 40 – *Em busca da consciência*. São Paulo, 2011.

MAYR, Ernst. *O impacto de Darwin no pensamento moderno*. Scientific American Brasil. Edição especial nº 17 – *Como nos tornamos humanos*. São Paulo, 2005.

MILTON, Katharine. *Alimentação e evolução dos primatas*. Scientific American Brasil. Edição especial nº 17 – *Como nos tornamos humanos*. São Paulo, 2005.

MOREIRA, Caio Margarido. *Não fui eu, foi meu cérebro!* Mente e Cérebro. Ano XVII, nº 211. São Paulo, 2010.

NEUWEILER, Gerhard. *A origem de nosso entendimento*. Scientific American Brasil. Edição especial nº 37 – *A ascensão do homem*. São Paulo, 2010.

NEVES, Walter A. *Pioneiros da América*. História viva. *A odisseia dos primeiros humanos*. Ano VI, nº 62, São Paulo, 2008.

_____; HUBBE, Mark. *Luzia e a saga dos primeiros americanos*. Scientific American Brasil. Edição especial nº 2 – *Novo olhar sobre a evolução humana*. São Paulo, nov. 2003.

ORNELAS, César Oscar. *Uma mão lava a outra*. Ciência & Vida. Edição especial, Ano II, nº 6 - Psique. Abr. 2012.

ORR, Allen H. *Sutilezas da seleção natural*. Scientific American Brasil. *A mais poderosa ideia da ciência*. São Paulo, Ano VII, edição especial nº 81. Fev. de 2009.

PICP, Pascal. *Os primeiros artesãos*. História Viva. Ano VI, nº 62 – *A odisseia dos primeiros humanos*. São Paulo, 2008.

POLLARD, Katherine S. *O que nos faz humanos*. Scientific American Brasil. Ano 7, nº 84 – *A ascensão do homem*. São Paulo, 2009.

Q. CHOI, Charles. *Uma teoria da fusão mortal*. Scientific American Brasil. Ano VII, edição especial nº 81 – *A mais poderosa ideia da ciência*. São Paulo, fev. de 2009.

SANTOS, Isabella Bertelli Cabral dos; e VARELLA, Marco Antônio Corrêa. *Quando a dor é a mãe*. Ciência & Vida. Edição especial, Ano II, nº 6 - Psique. São Paulo, abr. de 2012.

SHUBIN, Neil H. *Falhas de Projeto*. Scientific American Brasil. Ano VII, edição especial nº 81 – *A mais poderosa ideia da ciência*. São Paulo, fev. de 2009.

SOUSA, Altay Alves Altino de. *Casar ou comprar uma bicicleta*. Ciência & Vida. Edição especial, Ano II, nº 6 - Psique. Abr. de 2012.

_____; e Guedes, Álvaro Costa Batista. *Procura-se*. Ciência & Vida. Edição especial, Ano II, nº 6 - Psique. Abr. de 2012.

_____. *Uma longa história*. Ciência & Vida. Edição Especial, Ano II, nº 6 - Psique. Abr. de 2012.

STIX, Gary. *O legado vivo de Darwin*. Scientific American Brasil. Ano VII, edição especial nº 81 – *A mais poderosa ideia da ciência*. São Paulo, fev. de 2009.

_____. *Pegadas nítidas de um passado distante*. Scientific American Brasil. Edição especial nº 37 – *A ascensão do homem*. São Paulo, 2010.

TATTERSALL, Ian. *Como nos tornamos humanos*. Scientific American Brasil. Edição especial nº 17 – *Como nos tornamos humanos*. São Paulo, 2005.

263

_____. *Não estávamos sozinhos*. Scientific American Brasil. Edição especial nº 2 – *Novo olhar sobre a evolução humana.* , São Paulo, nov. 2003.

_____. *Partindo da África*. Scientific American Brasil. Edição especial nº 2 – *Novo olhar sobre a evolução humana*. São Paulo, nov. 2003.

THORNE, Alan G.; WOLPOFF, Milford H. *A evolução multirregional dos humanos*. Scientific American Brasil. Edição especial nº 2 – *Novo olhar sobre a evolução humana*. São Paulo, nov. 2003.

TSIEN, Joe Z. *O código da memória*. Scientific American Brasil. Edição especial nº 40 – *Em busca da consciência*. São Paulo, 2011.

VAN SCHAIK, Carel. *Por que alguns animais são tão inteligentes*. Scientific American Brasil. Edição especial nº 17 – *Como nos tornamos humanos*. São Paulo, 2005.

VARELLA, Marco Antônio Corrêa. *A era dos mal-entendidos*. Ciência & Vida. Edição Especial, Ano II, nº 6 - *Psique*. Abr. 2012.

_____. *Múltiplos parceiros*. Ciência & Vida. Edição Especial, Ano II, nº 6 - *Psique*. Abr. 2012.

VIEGAS, Lia Matos. *Que olhos grandes você tem*. Ciência & Vida. Edição especial, Ano II, nº 6 - *Psique*. Abr. 2012.

WARD, Peter. *Que futuro espera pelo Homo Sapiens*. Scientific American Brasil. Ano VII, edição especial nº 81 – *A mais poderosa ideia da ciência*. São Paulo, fev de 2009.

WHITE, Tim D. *Éramos canibais*. Scientific American Brasil. Edição especial nº 2 – *Novo olhar sobre a evolução humana*. São Paulo, nov. 2003.

WONG, Kate. *A filha de Lucy*. Scientific American Brasil. Edição especial nº 37 – *A ascensão do homem*. São Paulo, 2010.

WONG, Kate. *Decodificando o mamute*. Scientific American Brasil. Ano VII, Edição especial nº 81 – *A mais poderosa ideia da ciência*. São Paulo, fev. de 2009.

_____. *Em busca do primeiro homem*. Scientific American Brasil. Edição especial nº 2 – *Novo olhar sobre a evolução humana*. São Paulo, nov. 2003.

_____. *Estrangeiros na nova terra*. Scientific American Brasil. Edição especial nº 17 – *Como nos tornamos humanos*. São Paulo, 2005.

_____. *Genealogia humana*. Ilustrações: Viktor Deak. Scientific American Brasil. Ano VII, edição especial nº 81 – *A mais poderosa ideia da ciência*. São Paulo, fev de 2009.

_____. *Genealogia humana*. Scientific American Brasil. Edição especial nº 37 – *A ascensão do homem*. São Paulo, 2010.

_____. *O crepúsculo do homem de Neandertal*. Scientific American Brasil. Ano 8, nº 88 – *O mistério dos Neandertais*. São Paulo, set. de 2009.

_____. *O despertar da mente moderna*. Scientific American Brasil. Edição especial nº 17 – *Como nos tornamos humanos*. São Paulo, 2005.

_____. *O encontro de uma nova espécie*. Scientific American Brasil. Edição especial nº 37 – *A ascensão do homem*. São Paulo, 2010.

_____. *O menor dos humanos*. Scientific American Brasil. Edição especial nº 17 – *Como nos tornamos humanos*. São Paulo, 2005.

_____. *Quem eram os Neandertais?* Scientific American Brasil. Edição especial nº 2 – *Novo olhar sobre a evolução humana*. São Paulo, nov. 2003.

WRIGHT, Karen. *Os tempos da nossa vida*. Scientific American Brasil. *Especiais Temáticos*. São Paulo, 2011.

ZIMMER, Carl. *Quem sou eu?* Scientific American Brasil. Edição especial nº 40 – *Em busca da consciência.* São Paulo, 2011.

Publications

MONTEIRO, Euder. *Paleoantropologia para iniciantes – um curso ilustrado sobre a evolução físico-biológica dos humanos.* Disponível em: <www.paleoantropologia.com.br>. Acesso em: mai./2011.

RUSHTON, J. Philippe. *Raça, evolução e comportamento: uma perspectiva de história de vida ("A life history perspective"),* 2. ed. Abreviada. Disponível em vários sites, dentre eles, Charles Darwin Research Institute. Disponível em: <www.charlesdarwinresearch.org>. Acesso em: dez./ 2010.

Videos

Como nos tornamos humanos. Produtor: Graham Townsley. Coordenador: Kalindi Corens. Produção: NOVA. Distribuição: São Paulo: Ediouro; Duetto Editorial, 2009. DVD 1 (53 min.), DVD 2 (53 min.), DVD 3 (53 min.). Áudio: Inglês. Legendas: português.

Evolução – a aventura da vida. Produtor da série: Miles Barton. Produtor executivo: Phil Dolling. Produção: BBC, 2005. Distribuição: São Paulo: Abril, 2005. DVD 1 (Filme 01 – 50 min., Filme 02 – 50 min. e Filme 03 – 50 min.); DVD 2 (Filme 03 – 49 min., Filme 04 – 48 min. e Extras – 24 min.). Áudio: Inglês. Legendas: Português.

Origens da vida – a evolução das espécies. Produtores executivos: Michael Rosenfeld e Keenan Smart. DVD 1 (Filme 1 – 53 min. e Filme 2 – 53 min.); DVD 2 (Filme 3 – 53 min. e Filme 4 – 53 min.); DVD 3 (Filme 5 – 53 min. e Filme 6 – 53 min.) e DVD 4 (Filme 7 – 53 min. e Filme 8 – 53 min.). Áudio: Inglês. Legendas: Português.

Vida – desafios da vida. Produtora: Martha Holmes. (60 min.). Filme 1 da série Vida, produzida pela BBC, exibida pela Discovery Chanel através da Sky. Áudio: Português.

Evolução – a incrível jornada da vida. Scientific American Brasil. Sob licença da Ediouro. Seguimento Duetto Editorial Ltda. DVD 1 (Episódio I: A perigosa ideia de Charles Darwin, 120 min); DVD II (Episódio 2: Grandes mutações, 60 min; Episódio III: Extinção, 60 min); DVD 3 (Episódio IV: A corrida das espécies, 60 min; Episódio V: O porquê do sexo, 60 min); DVD 4 (Episódio VI: O Big Bang da mente, 60 min; Episódio VII: Ciência e religião, 60 min). Áudio: Inglês. Legendas: Português.

Fronteiras da Física – o universo elegante. Baseado no best-seller de Brian Greene, físico e matemático da Universidade de Colúmbia. Scientifc American. Duetto. DVD 1 (O sonho de Einsten; A corda é a base). 110 min. Áudio: Inglês. Legendas: Português.

Mistérios da mente. Apresentação: Professor Robert Winston. Produção e direção: Diana Hill. BBC. DVD 1: Mecanismos da inteligência, 50 min. Áudio: Inglês. Legendas: Português.

O corpo humano. Produzido por Richard Dale, Emma De'ath, Andrew Thompson, Peter Georgi, Christopher Spencer, Liesel Evans e John Groom. BBC. DVD 1 (Da vida real); DVD 2 (O milagre da vida); DVD 3 (O primeiro passo); DVD 4 (Da larva à borboleta); DVD 5 (O poder do cérebro); DVD 6 (Enquanto o tempo passa); DVD 7 (O final da vida); DVD 8 (Assim se fez). Tempo aproximado de cada DVD: 50 min. Áudio: Inglês. Legendas: Português.

O corpo sobre-humano. Produzido por Richard Dale, Emma De'ath, Andrew Thompson, Peter Georgi, Christopher Spencer, Liesel Evans e John Groom. BBC. DVD 1 (Transplantes); DVD 2 (Trauma); DVD 3 (Regeneração); DVD 4 (O inimigo interior); DVD 5 (De carrasco a defensores); DVD 6 (Tecno-genética). Tempo aproximado de cada DVD: 50 min. Áudio: Inglês. Legendas: Português.

A construção do ser humano. Produzido por Richard Dale, Emma De'ath, Andrew Thompson, Peter Georgi, Christopher Spencer, Liesel Evans e John Groom. BBC. DVD 1 (DNA); DVD 2 (Criação); DVD 3 (O segredo do sexo); DVD 4 (Juventude eterna). Tempo aproximado de cada DVD: 50 min. Áudio: Inglês. Legendas: Português.

América do Sul selvagem. Produzido e dirigido por Karen Bass. Fotografia de Barrie Britton, John Brown, Nick Gordon, Paul Johnson, Alastair McEwen, Rick Rosenthal, Stephen de Vere e Marck Iates. BBC. DVD 1 (Mundos perdidos); DVD 2 (O poderoso Rio Amazonas); DVD 3 (Grandes planícies); DVD 4 (Os Andes); DVD 5 (A floresta amazônica); DVD 6 (Praias de Pinguins); DVD 8 (Capivaras); DVD 9 (Piranhas); DVD 10 (A tropa). Observação: não foi possível adquirir o DVD 7. Tempo aproximado de cada DVD: 50 min. Áudio: Inglês. Legendas: Português.

Fronteiras da Física – o universo elegante. Duetto Editorial. Scientific American. DVD 2 (Bem-vindo ao universo de 11 dimensões, 60 min). Áudio: Inglês. Legendas: Português.

Maias – apogeu e ruína de um povo. Produção: Steven Talley. Narração: Salvatore F. Vecchio. Editora Abril. National Geographic. 50 min. Áudio: Inglês. Legendas: Português.

Egito – redescobrindo o mundo perdido. Direção: Fernando Fairfax. Narração: Andrew Sache. DVD 1 (Episódios 1, 2 e 3). 150 min. Áudio: Inglês. Legendas: Português.

Einstein revelado – o homem por trás do gênio. Einstein interpretado por Andrew Sache. Narração: F. Murray Abraham. Nova Internacional. DVD 1 (50 min); DVD 2 (50 min). Áudio: Inglês. Legendas: Português.

As grandes conquistas da ciência. Descobertas que transformaram o mundo em 100 anos de Prêmio Nobel. DVD 1 (50 min). Áudio: Inglês. Legendas: Português.

Os grandes avanços da medicina. Descobertas que transformaram o mundo em 100 anos de Prêmio Nobel. DVD 1 (50 min). Áudio: Inglês. Legendas: Português.

Caminhando com os dinossauros. Produtor da série: Tim Haines. Produção: Jasper James. Narrador: Kenneth Branagh. BBC. DVD 1 (Episódio 1: Sangue novo; Episódio 2: Tempo de Titãs; Episódio 3: Mar cruel). 90 min. DVD 2 (Episódio 1: O gigante dos céus; Episódio 2: Os espíritos da floresta de gelo; Episódio 3: A morte de uma dinastia). 90 min. Áudio: Inglês. Legendas: Português.

Origens. Duetto. Nova. DVD 1 (Vida fora da Terra). 52 min. DVD 2 (O nascimento do universo). 53 min. Áudio: Inglês. Legendas: Português.

Note: The year of publication of some videos is not quoted because it was not stated in the product documentation, perhaps for commercial reasons. In some magazines, I did not find the date of the publication and I chose to use a date based on the magazine's own articles. In some cases, I relate details of the articles of an entire journal when it would suffice to quote the journal covering them, but I preferred to quote them one by one in order to honor their authors.

Documentaries

In addition to these videos in DVD format, I watched and recorded several documentaries through the channels broadcast by SKY Brasil Serviços Ltda, particularly National Geographic, The Discovery Channel, and The History Channel. Unfortunately, due to a technical problem, I lost these recordings, and I was unable to cite information, such as production, direction, narration, etc. Nevertheless, for the sake of mere information, I list the titles of most of these documentaries, lamenting the omission of approximately a hundred others, of which I had not even noted the title, when a technical problem erased the recordings. In general, they last for 60 minutes and are dubbed into Portuguese. A few, however, have audio in English and have Portuguese subtitles.

The Science of Sex Appeal; The Great Family Tree; The Incredible Human Machine; The Mummy of Salta; The Biotechnology Revolution; The Revolution of Intelligence; The Quantum Revolution; Animal Anatomy – The Whale; Animal Autopsy - Crocodile / Giraffe; Kangaroos - The Dark Side; Incredible Brains - Memory; Violent Chimpanzees; Ice Dinosaurs; Immortality; Anabolics; Animal Attraction; Armies of Lions; Galápagos - Mutation; Galapagos 1, 2 and 3; Swine Flu; Sea lions; Wolves: the Predator and the Prey; Smart Monkeys; Japanese Monkeys; Deadly Creatures of the Desert; The White Sturgeon; What Darwin Never Knew; Prehistoric Predators; Human Prehistory - America; Sequoias - Giant Trees; Animal Sex; Triumphs of Life - Evolution; Triumphs of Life - Procreation; The Prehistoric Giant shark; Link 1 - The Story of Ida; Link 2 - The Story of Ida; The Future World - Superhuman; The Future World - Intelligence; Asteroid - 64 Million Years; Monkeys of Gibraltar; Striped Mutes; Monkey Men; The Biological Clock; Extraterrestrial Life; Early America 1; The Funeral of Tutankhamen; The Human Brain - Disc 1; The Human Brain - Disc 2; Prehistoric Monsters Disc 1; Prehistoric Monsters Disc 2; Evolution - Communication; Evolution - Eyes; Evolution - Poisons; Evolution - Flight; Evolution - Sex; Evolution - Jaws; Evolution - Size; Evolution - Forms / Communication; Evolution - Skin / Eyes; Life - Life Challenges; Life 1 - Reptiles and Amphibians; Life 2 - Mammals; Life 3 - Fish; Life 4 - Birds; Life 5 - Insects; Life 6 - Hunters and Hunts; Life 7 - Creatures of the Deep; Life 8 - Plants; Life 9 - Primates; Legacies of Egypt 1 - Religion and the Devil; Legacies of Egypt 2 - Cemeteries and Mummies; Legacies of Egypt 3 - Urban Centers; Legacies of Egypt 4 - Violence; Legacies of Egypt 5 - Pyramids; Legacies of Egypt 6 - Relics; The Animal Brain; Extreme Animals - Tools; Man Vs. Animal - Language; Man Vs. Animal - Medicine; Man Vs. Animal - Politics; Man Vs. Animal - Adoption; Man Vs. Animal - Tools; Building the Planet Earth; The History of the World: Revolution; The History of the World: The Age of Plunder; The History of the World: Under the Light; The History of the World: The Word and the Sword; The History of the World: The Age of Empire; The Incredible Human Machine; The Lost Tomb of Genghis Khan; Egypt - The World of the Dead; Naica - The Cave of Crystals; New Origins of Civilization; The Missing Link; Lost Civilization; Humanity - The Story of us All: Inventors;

Humanity - The History of all of Us: Revolutions; Humanity - The Story of All of Us: Treasure; Humanity - The Story of Us All: The New World; Humanity - The History of All of Us: Empires; Humanity - The History of All of Us: New Frontiers; Humanity - The Story of All of Us: Speed; Humanity - The History of All of Us: Pioneers; Humanity - The Story of All of Us: Survivors; Humanity - The Story of Us All: The Plague; Humanity - The Story of All of Us: Warriors; Humanity - The Story of Us All: Men of Iron; As the Earth Made Man; Tribal Fishing; I, Human; The Death Race; Mutant Planet; Animal Time - Men and Dogs, a Love for All of Life; Human Journey; In Darwin's Garden; The Barbarians; The Atlantic Forest and Life Cycles; The Hunter - Deadly Darts; The Hunter: Millenary Tactics; The Hunter: Bushmen of the Kalahari; The Hunter: Bow, Arrow and Smoke; Triumphs of Life: Animal Society; Life's Triumphs: The Power of the Brain; Triumphs of Life: Procreation; Animal intelligence; Extinctions; Mountain Gorilla - Episode 1 and 2; Creation - Charles Darwin; Troy; Ulysses; South Pacific; Inca Exploration; Chimpanzees - One Step Behind Us; The Gene of Violence; How Our Planet was Born: Mount St. Helens; How Our Planet Was Born: the American Ice Age; Earth's Most Lethal Eruption; Wild Indonesia; Megapharmotor 10.0; Man and Animal: Baboons in Survival; New Origins of Civilization; Sex in the Stone Age; The Two-Million-Year-Old Boy; The European Continent; Natural World; Smart Monkeys.

Internet

The initial intention was to share here the addresses of the main websites in which I researched. However, as I began checking the addresses, I saw something that I had already suspected: most of them no longer existed and others were so different that quoting them would be of no use to readers. On the contrary, it could be confusing because even in those that I found, the webpages were in other places of the same site and/or with totally different formatting. Therefore, I prefer to register here the great importance that the Internet had for the final result of this book and to thank all those who produced, wrote, organized, edited, designed, and directed all forms of communication to which I had access in my studies and in the preparation of this book.

ANOTAÇÕES

Made in the USA
Lexington, KY
18 July 2018